The
Mating and Breeding
of Poultry

HEN AND CHICKS

The
Mating and Breeding
of Poultry

Harry M. Lamon

Rob R. Slocum

THE LYONS PRESS
Guilford, Connecticut
An imprint of The Globe Pequot Press

The Lyons Press is an imprint of The Globe Pequot Press.

Printed in the United States of America

10 9 8 7 6 5 4 3 2 1

ISBN 1-58574-814-5

Library of Congress Cataloging-in-Publication data is available
on file.

Dedicated

to

Standard Bred Poultry

which has had such a pro-
found beneficial influence on
the great poultry industry
of the United States.

FOREWORD

TO the man or woman who desires some occupation or activity for a few spare hours as a relief from the fatigue, the strain and the routine of the regular day's work, poultry breeding, whether for the production of exhibition stock or of stock with heavy egg producing ability, is ideal. An intimate understanding of the laws or principles of breeding as well as of their application is necessary for success and poultry breeding is therefore an occupation which demands deep and discriminating thought and whose complex problems are at once a challenge and a stimulation to the intelligence of the breeder as well as being of absorbing interest.

Moreover, the poultry breeder finds opportunity for the expression of his artistic instincts in molding the form or shape of the birds and the color or combination of colors to meet his ideal. In addition, there is the gratification of the sporting instinct, the pleasurable excitement of competition in exhibiting the choicest specimens at the shows and the satisfaction which comes from a win in the realization that the breeder has surpassed the efforts of his competitors and produced birds superior to theirs.

Let no one to whom poultry breeding appeals hesitate to engage in it on the ground that he has not the room or facilities to enable him to compete with others more favorably situated. With only a back yard or village lot, and with the crudest equipment, it is possible to produce fowls of the highest excellence. Many winning birds in the leading exhibitions are produced by men and women who breed on a very limited scale without farms available for their poultry operations.

Poultry shows or exhibitions play a most important part in improving the different breeds and varieties. Not only do they serve to foster competition and create rivalry, thereby increasing interest in breeding, but they also make it possible to compare results. In this way the different breeders have an opportunity to see what other breeders have accomplished and to observe where their birds are strong or weak in comparison. Thus, they learn where they must seek to improve if they expect to work their way to the top or if they expect to stay there once they have arrived.

INTRODUCTION

THE publication here presented represents in its preparation the work of the authors extending over a period of about four years. It was inspired by the apparent need of some such guide for those who are beginners in breeding standard bred poultry or those whose experience is not as wide as is that of the breeders whose methods of mating are herein set forth.

It must not be misconceived that this publication is designed to or can in any sense replace the American Standard of Perfection. The Standard of Perfection describes in detail the ideal birds of the different breeds and varieties, and it is perfectly apparent that in order to insure success in attaining these ideals, the first thing to do is to purchase a copy of the Standard and to thoroughly familiarize oneself with the requirements of the ideal birds so as to have a clear-cut conception in mind of the goal which the breeder desires to reach as nearly as possible. This book is designed to supplement the Standard and to indicate the methods of breeding and mating which will be most likely to produce the results described in the Standard. In doing this, matings as used by many of the foremost and most successful breeders are given, and in addition, the tendencies or defects which are most likely to or which may prove troublesome are indicated and emphasized. In using this book, therefore, it is also necessary to make constant use of and reference to the American Standard of Perfection.

The matings described are those which breeders desire to use if the proper birds are available. It must be kept in mind, however, that it is never possible to secure perfect birds which possess exactly the characteristics desired and the problem often resolves itself into making the best use possible of the birds which the breeder has on hand. In doing this the aim should be to approach the matings as given as closely as possible and to bear in mind the necessity of offsetting any defects in birds which it may be necessary to use by strength in these particulars in the birds of the opposite sex in the mating.

The material contained in this book has been obtained from a wide number of sources. In the main it represents the practice of many of the foremost breeders of the country, together with the information which the authors possess on the subject. It will be noted that in some varieties separate and distinct methods of mating are described which in some cases may seem to be contradictory. These reflect the different opinions and methods of the breeders consulted who may be dealing with different blood lines having distinctly different breeding tendencies or who may be able to secure the results desired by different methods.

The authors wish to take advantage of this opportunity to acknowledge the splendid spirit of co-operation and cordiality of the following breeders in furnishing information and the inestimable aid which this information has been in preparing the book:

Newton Adams	W. H. Card
Adolph E. Anderson	A. Q. Carter
Wm. Anderson	Carl J. Carter
W. B. Atherton	Walter J. Coates
Balch & Brown	F. G. Cook
M. S. Barker	Newton Cosh
Frank G. Bean	Frank Davey
J. Y. Bicknell	Maurice F. Delano
L. H. Brown	L. J. Demberger
C. H. Brundage	Robert H. Essex
C. S. Byers	U. R. Fishel

James Glasgow
D. M. Green
W. J. Greenman
George W. Hackett
H. B. Hark
A. L. Hathaway
A. C. Hawkins
Wm. A. Hendrickson
S. G. Hoke & Sons
H. I. Hope
J. R. Huddleston
A. P. Ingraham
Paul Ives
M. R. Jacobus
John D. Jaquins
John C. Kriner
D. G. Kyler
C. H. Latham
Alfred Holmes Lewis
Thos. Lockwood
J. F. McKay
Cecil Manors
John S. Martin
Stanley Mason
George C. Meier
Richard Oke
Allen G. Oliver
D. L. Orr
Charles Pape
Frank L. Platt

J. M. Priske
Clyde H. Proper
Archie Rawnsley
Harold Rawnsley
Len Rawnsley
F. W. Rogers
R. A. Rowan
A. O. Schilling
Charles H. Shaylor
J. W. Shaw
H. C. Sheppard
Eugene Sites
Arthur C. Smith
Charles M. Smith
Courtland H. Smith
Edward F. Smith
W. A. Smith
A. & E. Tarbox
Lewis C. Taylor
E. B. Thompson
Lester Tompkins
H. V. Tormahlen
Arthur Trethaway
George W. Weed
Watson Westfall
R. J. Williams
Ralph Woodward
D. W. Young
Walter Young

The authors further wish to acknowledge their deep indebtedness to the following men of the Animal Husbandry Division, Bureau of Animal Industry, U. S. Department of Agriculture, for their critical review of the work and for their many helpful suggestions:

Alfred R. Lee, Animal Husbandman in Poultry Investigations.

J. W. Kinghorne, Junior Animal Husbandman in Poultry Investigations.

D. Lincoln Orr, Extension Poultry Husbandman.

D. M. Green, Extension Poultry Husbandman.

J. P. Quinn, Extension Poultry Husbandman.

Dr. Sewall Wright, Senior Animal Husbandman in Animal Genetics.

Still further acknowledgment is due Mr. J. W. Kinghorne, who has so painstakingly and thoroughly indexed the contents of this book and who has contributed some illustrative drawings.

If the authors have in even a slight degree, accomplished the purpose which they had in mind in writing this book, namely, to assist beginners or breeders to secure greater success in breeding high class exhibition stock or stock of high egg laying ability, or if they have contributed to bring about these results in a shorter time and thereby have contributed to the advancement of standard bred poultry, they will feel amply repaid for the great amount of work which has gone into its preparation.

DEFINITION OF COMMON BREEDING TERMS

NOTE—For the definition of terms not included in this list, see the American Standard of Perfection.

ANGULAR APPEARANCE—The lack of smoothness in the appearance or outline of fowls caused by the abrupt junction of certain sections so as to form an angle instead of a curve.

BEEFY COMB—A comb that is large, thick and of more substance than called for by the Standard.

BEETLE GREEN—A metallic green of considerable luster, similar to the color found on a beetle's back.

BLOOD LINES—Refers to the breeding or admixture of blood which a fowl possesses. Definite blood lines are the result of a fixed plan of breeding with the consequent concentration to a greater or less degree of the blood of some particular individual, family or strain. Well-established blood lines are characterized by a greater uniformity of the fowls belonging to these lines with respect to some character or set of characters.

BOOTS—Term used to designate the shank feathering of some of the varieties of Bantams.

BREAST, CROOKED—The breast or keel bone deformed by being notched, bent, or turned to one side usually due to allowing the birds to roost when too young or to an injury.

BUCKLED HOCK JOINTS—A term applied to hock joints especially in Games, which are weak and do not enable the bird to stand erect.

BULL HEAD—A broad, thick head which when found in exhibition Games is a serious defect.

COARSE COMB—One that is large, thick and rough in texture.

COCKEREL MATING—A mating made for the special purpose of producing exhibition males when double mating is employed.

COMBINATION MATING—A mating where one male is used on two different types of females.

CONSTITUTIONAL VIGOR—A combination of those qualities which make for health, strength and vitality in fowls.

COW HOCKED LEGS—Legs which curve or bend back at the hock joint instead of being straight under the fowl.

xiii

CROOKED TOES—Referring to toes which are other than straight.

CROOKED BACK—A deformity of the back often the result of an injury which causes it to be turned or twisted or unevenly developed on one side.

CROSSING—A mating of fowls of two separate strains, varieties or breeds.

CROW HEADED—A long, drawn-out head and beak resembling that of a crow.

DEFECT—An imperfection or undesirable quality found in a bird which should be guarded against in selecting exhibition and breeding birds.

DOUBLE MATING—The use of two separate and distinct matings within a single variety, one of which is especially designed to produce exhibition males and the other exhibition females.

DUCK SHAPED TAIL—A broad tail somewhat horizontally spread like a duck's tail.

FLAT SHANKS—A defect found in some breeds that should have round shanks. The flat portion usually extends up the front of the shank.

FLAT WINGS—Wings which when folded present a broad, flat surface instead of being curved or bowed to conform to the shape of the body.

GAMY TAIL—See Whip tail.

GOOSE NECK—A defective neck formation resembling that of a goose.

GYPSY FACED—Dark purple color of face as found in the Brown Red Game, Black Sumatra and Silkies.

PATCHINESS—The presence of irregular spots or patches due to different shades of color or to the uneven distribution or variation in the size of the markings in the feathers of an individual.

PINCHED HEAD—Elongated narrow head.

PINCHED TAIL—A narrow tail due to the main tail feathers being closely folded instead of well spread.

PULLET MATING—A mating made for the special purpose of producing exhibition females when double mating is employed.

PURPLE BARRING—A barring of iridescent purple extending across the feathers and more or less commonly found in the plumage of black varieties or of the black sections of other varieties.

ROACH BACK—A term usually used as synonymous with crooked back. Also used occasionally to denote a curved or arched back.

RUSTINESS—A reddish tinge appearing on the surface feathers.

SHEEN—As commonly used indicates the lustrous greenish or purple color occurring on black feathers.

SINGLE MATING—A mating made for the purpose of producing exhibition birds of both sexes.

SMOKY—An admixture of dark or black in the ground color of feathers, causing a general smudgy or dirty effect as though the feathers had been exposed to smoke. Smokiness usually appears in the main wing and tail feathers.

SMUTTY—When black, dark or slate of varying intensity occurs in the under color of breeds which should be free from it, the under color is said to be smutty as in the Rhode Island Red. In breeds having plumage which is particolored and one of the colors is black or dark, some of the dark color often extends over into the lighter colored portion of the feathers, destroying its clearness and causing it to appear dirty or dingy. It is then said to be smutty as in the Barred Rock.

SPLIT CREST—One that is parted or divided. A defect in crested breeds.

SPLIT TAIL—One in which the main tail feathers are divided, leaving an open space in the center, through which the sickle feathers may and often do fall.

SPLIT WING—Also known as slipped wing, referring to one that is not properly held in place, but which hangs down partly unfolded.

STAMINA—A term used in connection with games indicating health, vigor and endurance.

STANDARD MATING—See Single Mating.

STILTINESS—The appearance of a bird being out of proportion due to too great a length of leg.

STUD MATING—Referring to the practice of keeping the male birds used in separate quarters or coops and bringing to them as they lay, the females with which they are to be bred.

TUFT—A group of feathers or a tendency toward a small crest appearing on the heads in some breeds.

"U" SHAPED BACK—A back which owing to the upright position of the tail gives the appearance of the letter "U" as in the Langshan or Seabright Bantam.

WHIP TAIL—A very narrow pinched tail ending in a point commonly used in referring to the tail of exhibition Games.

CONTENTS

xvii

CHAPTER IX

CHAPTER X

CHAPTER XI

CHAPTER XII

CHAPTER XIII

CHAPTER XIV

CHAPTER XV

The Silkies—The Sultans—The Frizzles.

CHAPTER XVI

Time of Hatching—Feeding and Management of Growing Stock—Feeding White Fowls—Value of Shade—Selecting the Birds to be Prepared for the Show—Cooping Birds for Training—Training—Danger of Over-Cooping—Feeding Birds Which Are Being Conditioned—Cleaning Shanks and Toes—Washing—Bluing White Birds—Bleaching and Cleaning—Other Care of Plumage—Shipping Birds to the Show—Care of Fowls in the Show—Treatment of Birds After the Show—Making Entries.

LIST OF ILLUSTRATIONS

Fig 1—Nomenclature Chart of a Male Fowl

CHAPTER I

PRINCIPLES OF BREEDING

This book deals with the different varieties of chickens described in the American Standard of Perfection and the principles of breeding in order to produce specimens of a high degree of excellence and obtain as great a proportion as possible of good quality. Breeding chickens with this object in view is a very old pursuit and engages the attention and interest of a great number of individuals. Were the problem of mating simply that of selecting individuals possessing the desired characteristics with the certainty that the offspring would be identical in character with the parents, it would be a comparatively simple one. As a matter of fact, it is not a simple question of like producing like, and for this reason the successful breeding of chickens of high quality is a difficult problem requiring the most painstaking care and study, and good results are obtained with a reasonable degree of certainty only as the result of long experience.

Before describing the matings which the experience of breeders has shown are most likely to give good results, it is desirable to consider briefly certain laws or principles of breeding which are more or less commonly met with in breeding and with the results of which, if not their theory, experienced breeders are familiar.

Like produces like.—This is a common expression employed by breeders. Within certain limits it is true. It means simply that there is usually a strong resemblance between parents and offspring, stronger than that between the offspring and unrelated fowls. Thus when two fowls of the same breed or variety, which are from stock which has had no foreign blood introduced for a considerable time, are

1

mated together, the offspring will all, or nearly all, resemble the parents quite closely in so far as the breed or the variety characteristics are concerned, and usually also with respect to some of the individual peculiarities of one or both of the parents. Pure-bred White Leghorns will produce offspring which are palpably White Leghorns both in type and color; Barred Plymouth Rocks will produce Barred Plymouth Rocks, etc. The offspring are likely also to inherit in some degree the shape of the parents, the comb peculiarities or other similar traits. It must be borne in mind, however, that it is impossible to select parents which are identical in their characteristics, and as each parent may and does have an influence upon the offspring, it can be easily understood why the offspring may not be identical with one or either of the parents. It must also be remembered that the grand-parents, and in turn their parents and the ancestors still further back, all exert some influence and tend to bring about variation.

Variation.—The differences which exist between the off-spring and their parents, or between offspring from the same mating, range from those rather slight in degree to those which are quite marked. For example, from a Rhode Island Red mating in which both parents are of fairly good color and free from smut in undercolor, there may be obtained offspring ranging from very light, poorly colored birds to dark, good colored birds, and from those with a clear red undercolor to those showing white or to those showing smut in undercolor. The offspring may therefore not only differ markedly from one another, but may also differ greatly from their parents, and may vary from those much poorer than the parent stock to those much better than the parent stock. It is the selection of those individuals showing variation in the desired direction that constitutes most of the endeavor in breeding. It is by the use of such individuals that breed-ers expect to make progress in their breeding operations. In the long run, this is usually a successful means of obtaining improvement, as the success of countless breeders will testify.

Sports or mutations.—Occasionally there occur individuals with characteristics which differ very markedly either in kind or in degree from those possessed by any of the

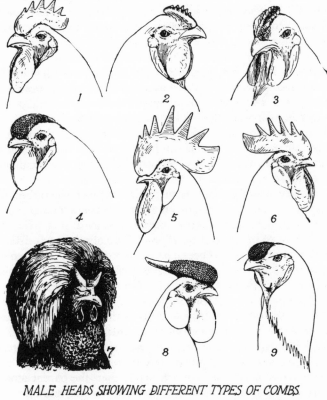

MALE HEADS SHOWING DIFFERENT TYPES OF COMBS

1 SINGLE	2 PEA	3 PEA
4 ROSE	5 SINGLE	6 SINGLE
7 \ SHAPED	8 ROSE	9 STRAWBERRY

Fig. 2. (From the Bureau of Animal Industry, United States Department of Agriculture.)

known ancestors. These individuals are known as sports or mutations. Usually there is a marked tendency for sports to breed true, and consequently if the sport happens to be of a desirable kind, it may result in much more rapid or much greater improvement than could have been obtained by the more gradual process of selecting less marked variations. The White Plymouth Rock is said to have arisen from white chicks produced from a Barred Plymouth Rock mating, and the former variety is commonly considered to be a sport from the latter.

Reversion or atavism.—By reversion or atavism is meant the appearance in the offspring of characteristics which do not exist in the immediate parents or ancestry, but which did exist in some of the ancestors several generations back. This is a common occurrence and is generally recognized and understood by breeders. Sometimes the reversion goes far back to some character in the distant ancestry, and this may be so far back that the breeder has no knowledge that such characters existed, with the result that the reappearance of the character puzzles him. An example of a common reversion is the continued reappearance of stubs on the shanks of Wyandottes and Plymouth Rocks, or of the down which occasionally appears between the toes of Leghorns, in spite of the use for generations of clean legged birds as breeders. The appearance of these rudimentary feathers undoubtedly is a reversion back to birds with feathered shanks which have been used at some distant time in making the breed. The more recently such birds have been used in the breeding, the more frequent and the more troublesome are such reversions. It is unwise to use individuals for breeding which show such a tendency toward reversion, particularly if the reversion is frequent in appearance.

Environment.—The conditions under which poultry grow and live may affect their development and may thus cause a modification in the size or type of the fowls. The environmental conditions which most commonly affect poultry are climate, feed and the various features of management, including housing, feeding and care of the eggs during incubation, and of the chicks during the growing period. Environmental conditions, particularly those of climate, may cause a modification of size or type, but affect the fowls, it is now believed, only in the way of preventing them from developing their inherited characteristics to the fullest extent, and do not interfere with the transmission of these. The effort should be made to control or modify environment in so far

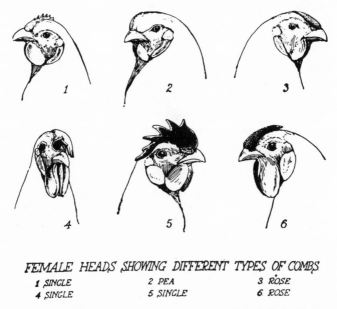

FEMALE HEADS SHOWING DIFFERENT TYPES OF COMBS

| 1 SINGLE | 2 PEA | 3 ROSE |
| 4 SINGLE | 5 SINGLE | 6 ROSE |

Fig. 3. (From the Bureau of Animal Industry, United States Department of Agriculture.)

as possible, so that it will be favorable for the full development of desired characters, such as high egg production, which are inherited by the fowls.

Inheritance of acquired characters.—The question arises whether or not characters which are due to environment, and which are commonly called acquired characters, are inheritable or transmissible. For example, is the greater size which a certain variety of chickens may reach, and which is due to climatic conditions, passed on to the offspring? It must be remembered, in considering this question of the inheritance of acquired characters, that in many cases the modification which may be observed in a race, while apparently due to the inheritance of characters modified by environment, is probably often due to the fact that individuals which do not show characters varying in this direction are unsuited for the particular environment under which they are living, and are therefore gradually eliminated by the process of natural selection, while the individuals varying in the direction in which the modification has occurred were especially well fitted for the environment under which they live. Under these conditions the modification which occurs will be largely if not entirely due to the process of natural selection, rather than to the inheritance of acquired characters.

In general, it may be said that unless environment has a modifying effect upon the germinal material of the individual concerned, the acquired character cannot be inherited. Experiments indicate that the germinal material is very difficult to affect in any way. It may be said that the weight of evidence is at present against the view that acquired characters are inherited.

In this connection it may be well to call attention to the matter of whether or not mutilations or deformities are inherited. Deformities due to faulty development, either embryonic or in the later life of the chick, or mutilations which may occur are very rarely if ever inherited. Thus we find that a fowl which has lost a toe is no more likely to produce offspring minus a toe than are those with all their toes intact. The long continued custom of game breeders in dubbing, or shearing off the combs of the males, has produced no appreciable tendency for the offspring to come without combs. It may be safely said, therefore, that poultry breeders need feel no concern over the possibility of mutilations and but little over deformities being inherited, except such as seem to be in the blood of the fowls, as evidenced by their frequent occurrence in the various generations.

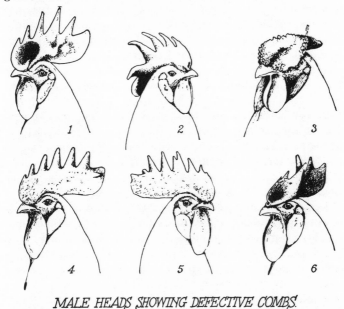

MALE HEADS SHOWING DEFECTIVE COMBS.

1 THUMB MARK.	2. LOPPED (SINGLE)	3. HOLLOW CENTER
4 SIDE SPRIG	5 UNEVEN SERRATIONS	6. TWISTED

Fig. 4. (From the Bureau of Animal Industry, United States Department of Agriculture.)

Regression.—By regression is meant the general tendency for the offspring of a mating to be, on the average, nearer the mean, or average for the race, than were the parents composing the mating. If the parents are selected individuals, well above the average with respect to any character or set of characters, the offspring will average poorer than the parents and be nearer the average for the race. On the other hand, if the parents are below the average for the race, the offspring still tend to be nearer the average than their parents, and therefore to be better than their parents. Since, in breeding poultry, the effort is always made to have the individuals composing the mating well above the average, it is the tendency toward a lower average in the offspring which particularly concerns the breeder. He must continuously fight this tendency to go back toward the average, and this must be done by continued selection.

It is this law of regresssion which makes it necessary for a breeder to continue to select with the greatest care, even after he has attained to a high degree of excellence in his flock, for just the minute he stops with the idea that he can rest on his laurels and maintain the degree of quality which he has secured, he will find that he is losing ground.

Correlation.—When it is observed that a certain character is likely to be associated with a certain other character, there is said to be a correlation between these characters. There is undoubtedly some general correlation of characters in poultry. Correlation is quite subject to breeding and selection, but when one desires to secure certain results in breeding, the characters which usually go together should be considered in selecting a type or standard for an ideal. Correlation may be of much value in fixing types, or, on the other hand, it may prove troublesome. It is usually true that males which are inclined to be too leggy are also inclined to be flat or lacking in breast. It also seems to be general observation that high egg production in any breed or variety is more likely to be associated with females which are not above the average size for the breed than with females which are much above the average size for the breed. A white shelled egg and white ear lobes are usually correlated, and breeds or individuals are rare which show good red ear lobes and which produce white shelled eggs.

Influence of the sire or dam on type and color.—The opinion is frequently held by breeders that the parents of one sex have more influence on type or shape, and those of the other sex upon color.

Unfortunately, the breeders are themselves divided as to which sex exerts the influence upon color, and which upon type. The fact seems to be that in general there is equal inheritance from the sexes in all characteristics. The real exceptions include a few characteristics of various kinds which are linked with sex in inheritance in such a way that there is no transmission from dam to daughters. These so-called sex-linked characters are discussed in a later section. Apparent exceptions may arise from differences in the pre-potency of individuals. It may happen in a particular mating that the male is prepotent in color and the female in type, but in another mating the situation may be reversed, or else some one parent may be prepotent in both respects. There is, in short, no general law connecting prepotency and sex.

MALES WITH DEFECTIVE TAIL CARRIAGE.
1 SQUIRREL   2 WRY

Fig. 5. (From the Bureau of Animal Industry, United States Department of Agriculture.)

Control of sex.—The control of the sex of offspring in chickens, as well as other domestic animals, is a question, the solution of which has long been sought. Many theories have been advanced and some breeders believe that they can control to some extent at least the sex of the offspring. The most commonly advanced theory among chicken breeders is the claim that a cockerel mated to hens will give a greater proportion of females than males, while a cock mated to pullets will give a greater proportion of males. This theory is doubtless nearly if not quite identical with that advanced by some breeders of other domestic animals, and seems to depend upon the idea that the more mature parent is sexually more vigorous than the less mature parent, with the result that more of the offspring will be of that sex. No experimental work or carefully kept records have as yet given good support to this idea and it is impossible to give it much credence. Indeed, at the present time it may be said with full assurance that no method has been found which will control at will the sex of the offspring in chickens.

It is, moreover, very doubtful whether any such method is possible. The mechanism by which sex is normally determined is now well understood, but unfortunately is a mechanism which seems to be beyond human control. In poultry, the males produce only one kind of germ cell, but the females produce two kinds in approximately equal numbers. One of these kinds develops into males, the other into females. The potentiality of an egg in this respect depends on the presence or absence of a small body (the X chromosome), which goes into the egg at a very early stage, or is discarded, wholly by chance.

Prepotency.—Prepotency is the word used to describe the ability of an individual to impress its characters or qualities upon its offspring. An individual may be termed prepotent either when it can transmit a certain character or set of characters to a considerable proportion of its offspring to an appreciable degree, or when it can transmit some character or set of characters to one or a few offspring to a very high degree. Thus we might speak of a White Wyandotte male of outstanding excellence of type as prepotent if he could transmit this excellence of type in a high degree to only one or two of his sons or daughters. Or we might speak of him as prepotent if he could transmit this excellence of type in a high degree, or even in a less high degree, to 50 or 60 per cent of his offspring. Again, we might speak of a Leghorn male as prepotent with respect to the character of egg production, if he were able to transmit to only one or two daughters the ability to lay 250 eggs, even though the rest of his daughters were below the average in that respect; or we might call him prepotent if he could transmit to 50 per cent or more of his daughters the ability to lay more than the average, say 180 eggs. The first of these instances may be said to represent prepotency in degree, and the second prepotency in numbers. Most poultry breeders in speaking of prepotency mean prepotency in numbers rather than prepotency in degree.

As can readily be seen, this quality of prepotency, when manifested in a desirable direction, is an extremely valuable one, and one of which the wise breeder takes full advantage. Once a bird, either male or female, has proven its prepotency, it should be used as a breeder just so long as it is in breeding condition. The only known method by which prepotency can be developed in a flock is through close breeding, a system which is discussed later.

Contamination.—The theory of contamination has never been so strongly or so widely held among poultrymen as among breeders of other animals, particularly dogs and horses, probably largely because the chick does not develop to an appreciable degree within the body of the mother. By contamination is meant that the service of a male upon a female, or the growth of an embryo within the female as the result of such service, has a lasting effect upon the female, so that succeeding offspring, although by different males, will bear a resemblance to the first or previous male or males. Such an influence or contamination, even if it existed, would have no perceptible or at least no important effect as long as the female and the different males were of the same variety in chickens or breed in the case of other animals. But if one of the males were of a different variety or breed, such contamination would immediately have an important and a disastrous effect. For were there such contamination, a White Leghorn hen, for example, which had been bred to a Silver Spangled Hamburg male would no longer have any value as a breeder of White Leghorn chickens, even when mated to White Leghorn males, since the chickens would show the influence of contamination due to the Hamburg male. The theory of contamination finds little support from poultry breeders, as there are no reliable authentic records to support it, while it is common practice to allow hens to run with males of a variety other than their own and to breed them subsequently to males of their own kind without the slightest bad effect. It is easy to account for the supposed observed cases of contamination by the accidental mating, without the breeders' knowledge, of the hen with another male than the one with which she is supposed to be mated.

Mendelism.—In the preceding sections, we have attempted to present the leading principles of breeding as they have been brought out by practical experience. It should be added that a great deal has been discovered in recent years about the laws of heredity, which underlie these practical principles. It would, however, go beyond the scope of this book to go into the subject in detail. There are a number of books on the subject which may be recommended to those who are interested. The following may be mentioned:

"Genetics in Relation to Agriculture," by E. B. Babcock and R. E. Clausen, 1918.

"Genetics and Eugenics," by W. E. Castle, 1916.

"Mechanism of Mendelian Heredity," by Morgan, Sturtevant, Muller and Bridges, 1915.

Fig. 6. Standard bred flock of Single Comb White Leghorns, showing uniformity of type and color. (Photograph from the Bureau of Animal Industry. United States Department of Agriculture.)

According to earlier views, heredity was looked upon as fluid in nature. We still use such expressions as half blood, quarter blood, etc., which convey this idea. It now appears, however, that the hereditary material is composed of units which are handed on almost indefinitely without change. Their behavior may be compared better to that of little solids contained in the germ cell than to fluids. The variations in most characteristics, it is true, depend on so many of these units that the mode of inheritance after crossing may still be compared roughly to the blending of fluids. A comparison of the hereditary basis of a characteristic to a pile of shot is better, however, when it comes to explaining the effects of inbreeding, the nature of prepotency, etc.

A few characteristics depend on such a small number of units that the effects of the different ones are easily followed in crosses. Mendel's law of heredity was originally discovered in these simple cases. It is now considered probable that Mendel's law is the general law for all heredity, but it is difficult to demonstrate except where the number of units is small, and hence is in most cases of little more than theoretical importance.

Certain cases of simple Mendelian inheritance frequently come under the eye of the poultryman who makes crosses. As an illustration, let us consider what happens when a fowl from a strain which produces rose combs only is mated with a single comb, which, as breeders know, breeds true. The offspring all have rose combs. Rose comb is said to be dominant over single comb. The crossbreds breed very differently, however, from their rose comb parent. They breed as if half of their germ cells (spermatozoa in the male, eggs in the females) transmit rose comb only, while half transmit single comb only. This is easily seen on crossing them with single combs, in which case they produce 50 per cent rose combs and 50 per cent single combs. The rose combs of this generation, when crossed with single combs again, produce 50 per cent rose combs and 50 per cent single

Fig. 7. A mongrel flock lacking in uniformity of type and color. (Photograph from the Bureau of Animal Industry. United States Department of Agriculture.)

combs, although the chicks are seven-eighths blood single comb. This system of mating can be carried on indefinitely, the rose comb chicks always producing 50 per cent rose combs and 50 per cent single combs, regardless of the amount of blood from single comb strains. When two of these crossbred rose combs of any generation are mated with each other, 25 per cent of the chicks will have single combs. The reason is easily seen. Half of the eggs transmit single comb and half of these, or one-quarter of all eggs, will, by chance, be fertilized by spermatozoa which transmit single comb, the result being that 25 per cent of the chicks fail to get rose comb from either parent and so are single combs. Conversely, it is easy to see that there will be another 25 per cent of the chicks which get rose comb from both parents and will breed like pure rose combs, although in blood they may be almost wholly of a single comb strain. The remaining 50 per cent of the chicks get rose comb from one parent, single comb from the other, and breed like the original crossbreds, that is, half their germ cells transmit rose comb and half single comb.

In this case, one of the opposed characteristics (rose comb) was dominant over the other (single comb). This is very often the case, but it is not a general rule. The Blue Andalusian fowl resembles the crossbred rose combs in that it produces two kinds of germ cells in equal numbers. A chick produced by the union of like germ cells from a pair of Blue Andalusians is either a splashed white or a black (which may be mixed with red or other colors, dependent on other hereditary units). Blues must receive the factor for black from one parent and that for white from another. Thus the mating of blue by blue produces 25 per cent black, 50 per cent blue and 25 per cent splashed white, no matter how much blue ancestry there is. The only way to obtain 100 per cent blue is to cross with each other the blacks and whites derived from Andalusian stock. It may be well to add that while twice as many blues are obtained by this method as by crossing blues with each other, the quality of blue may be expected to be poorer because of working in the dark in selecting the parents.

In the ordinary breeding operations of the man dealing with a pure breed or variety, Mendel's law will have little practical application. It is in crossing for the purpose of securing new combinations of characters that it may have a place. The fact that some characters may be crossed without blending and can be extracted again in a pure state in a short time is of value in securing the new combination of characters desired. Most breeders, however, find this law difficult to understand and confusing, and make little conscious use of it in their breeding operations. Moreover, relatively few characters seem to behave as typically unit characters, and if their behavior is Mendelian it is probably obscured by the fact that they are not simple unit characters, but are complex characters in themselves, more than one factor being involved in their make-up and reacting possibly in different directions with results which are not Mendelian in a sharp, clear-cut manner in so far as their interpretation is concerned by the average individual.

Sex-linked inheritance.—It has already been noted that in poultry the females produce two kinds of eggs, those which can develop into males and those which can develop into females, while males produce only one kind of germ cell. It happens that certain hereditary factors are linked with sex in such a way that females cannot transmit them in the eggs which are to develop into females. The theory is that the female-determining eggs are such because they lack a certain material body which is present in the male-determining eggs. Hereditary units which are transmitted in this same material body may or may not be present in the germ cells of males or in the male-determining eggs of the females, in all of which the material body is present, but they can never be transmitted in the female-determining eggs, in which the material body is absent. For example, the Barred

Plymouth Rock male can transmit barring both to his male and female offspring, while the Barred Plymouth Rock female is limited in transmitting barring to her male off-spring only. This is, of course, not apparent when a Barred Plymouth Rock male and female are mated together, for the offspring of both sexes inherit barring from the male parent. When, however, the Barred Plymouth Rock is mated with a non-barred variety, it immediately becomes apparent. If a Barred Plymouth Rock male is mated with Dark Cornish females, all the resulting offspring, both male and female, will be barred. Since this barring could not come from the females, it must have been inherited by both sexes from the male parent. If, on the other hand, Barred Plymouth Rock females are mated with a Dark Cornish male, only the male offspring will be barred. It is apparent, therefore, that the Barred Plymouth Rock females could transmit barring to the male offspring only and not to the female offspring.

CHAPTER II

PRACTICES OF BREEDING

Breeding tendencies.—Different strains or stocks of the same variety may and usually do have different breeding tendencies. Just what these tendencies are can be determined only by an intimate acquaintance with and close study of the stock in question, as the result of breeding for some time. Such a knowledge is most essential for success, and the breeder who fails to study his stock for this purpose will not attain the highest success. Knowing the breeding tendencies enables the breeder to count upon those tendencies which are strong in the right direction and at the same time enables him to take especial pains to guard against undesirable tendencies in making up his matings. Knowledge of the breeding tendencies will often have an important bearing on the selection of individual birds to be used in a mating and renders the selection of the individuals as breeders much less of a matter of guesswork. For example, if a flock of White Plymouth Rocks seldom produces individuals showing slipped wings, it might be desirable and wise to use a bird showing this defect if he possessed certain other qualifications of an outstanding character, whereas if the flock showed a tendency toward slipped wings it would be unwise to use such a bird.

It is pretty generally recognized that even an experienced breeder may purchase the best stock obtainable of any variety and still be unable to keep the quality of the stock up to that attained by its former owner, at least for some years. This is due to the fact that he is unfamiliar with its breeding tendencies and must breed the stock for several years before learning just what its breeding tendencies are.

Selection for vigor.—The vigor of the stock must be kept up. If this is not done, bad results will be obtained in fertility, hatchability and the rearing of the young stock. Success, whether in utility or exhibition stock, or both, depends upon the ability to keep the flock healthy and upon the ability of the birds to breed, of the eggs to hatch and of the chicks to be raised and to make good growth. Selection for vigor is, therefore, most important; it is fundamental. The breeder must not be blinded by excellence of color to the necessity for vigor, strength and constitution. It is not always possible to select the most vigorous birds, but the appearance of the fowls and their actions are a pretty reliable index to their health, vigor and condition. Fowls which show good, bright color of eye and head parts, glossy, clean plumage, fairly good substance of bone, and which are active and sprightly in carriage and have made good, continuous growth are usually strong, vigorous individuals. Avoid fowls which show sunken eyes, long, narrow, snaky heads, toes long for the breed, and which are listless and inactive, and whose growth has been slow, uneven or unbalanced.

Fig. 8. Barred Plymouth Rock female of low vigor showing slipped wing and crow head. (Photograph from the Bureau of Animal Industry, United States Department of Agriculture.)

Males, especially those over one year old, sometimes become weak in their hock joints. They show this by standing unsteadily on their legs and raise their feet unusually high in stepping. Males which are too long in leg seem to be more susceptible to this condition. Such males should not be used as breeders, for they are not likely to fertilize many eggs.

Inbreeding.—Inbreeding is the mating of related individuals. It is a practice commonly used by breeders for the purpose of concentrating the blood of individuals or families which show the desired characters or quality. It may be said with certainty that practically no outstanding success in breeding has been made without a considerable amount of inbreeding. The great strides which have been made in the improvement of breeds or varieties are due to such intensification or purification of the blood of outstanding or especially valuable individuals or families of the breed or variety. Such intensification of blood aims at and results in making the characteristics of the individuals or families so used breed more true.

But while inbreeding results in the intensification of the good qualities of the individuals used, it also results in the intensification of all the characters of such individuals. It is here that the danger of inbreeding lies, for if there happens to be, as there usually are, bad characters or weaknesses in the individuals, these are intensified as well. For this reason where inbreeding is practiced, too great care cannot be exercised in the selection of the individuals used, to see that they have no particularly bad traits, and to make sure that they are vigorous and of strong constitution. Without such care in the selection of the individuals inbreeding is in the end almost sure to result in failure. The closest forms of inbreeding, such as son on mother, daughter on father or full brother and sister, are sometimes called **in and in breeding.**

Where inbreeding is practiced the greatest watch-
fulness should be exercised to discover the first signs of
any weakness or bad effects. If these appear it is then
time to abandon, in some of the matings at least, such
close inbreeding as is being practiced and to attempt the
cautious introduction of new blood.

As an example of what can be done in inbreeding
without detriment to the stock, may be cited the case of
a breeder who, starting with a trio of birds, has bred for
thirteen years without the introduction of a drop of new
blood. During this period he has improved size, egg
production and color and has seen no weaknesses develop which can be attributed to inbreeding.

Line breeding.—Line breeding is a form of systematic
inbreeding in which an effort is made to keep away from
too close inbreeding. It is really in its ordinary use,
breeding confined to the blood lines of a single family.
The details of this practice vary considerably with the
different breeders, but the purpose is the same in each
case, namely, to avoid the necessity of introducing blood
of another strain or family with the disastrous results to
the uniformity of the strain which often accompanies
such an introduction of blood. Line breeding, whether
known by this name or not, is almost universally used by
successful breeders, but is often accompanied by the
occasional and judicious introduction of outside blood.
The chart following, Fig. 9, shows one method of
line breeding. In this chart, Circle 1 represents a male
and Circle 2 a female, whose blood it is desired to use in
the line breeding without the introduction or at any rate
without the frequent introduction of new blood. Male 1
mated with female 2 gives offspring 3, which are one-half
each of the blood of the male and the female. A male
from 3 mated with the original female 2 gives offspring 6,
which are three-quarters of the blood of the female. Similarly, a female from 3 mated with the original male 1

gives offspring 4, which are three-quarters of the blood
of the male.   A male from 7 mated with a female from 4
gives offspring 9, which are 13-16ths of the blood of the
male.   Further, a male from 9 mated with a female from
7 gives offspring 13, which are 27-32nds of the blood of

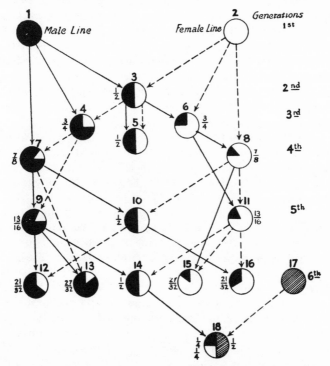

Fig. 9—Line Breeding Chart—The circles represent individuals or groups of
individuals used in the breeding or resulting from the matings made.   Black
represents the male line and white the female line, the relative amount of
each color in a circle as well as the fractional figures at the side indicating the
relative amount of each blood carried.   The large number at the top of each
circle is given to it so as to make the discussion of the matings in the text
clear and easy to follow.   A solid black line connecting one circle with another indicates that a male from the first group was used in the mating to
produce the second group, while a dotted line indicates that a female was
used.   The shaded circle at the lower right-hand corner indicates the introduction of a third blood line, and circle 18 shows the proportion of the three
bloods in the resulting offspring.   (After Felch, Pierce and Lippincott.)

the male. As will be apparent from the chart, various other matings are possible and even some which are not shown, all of which would have the effect of producing offspring with varying preponderance of the male blood. Exactly similar matings are possible as shown which will result in the same preponderance of the blood of the female. At any time, by mating individuals from groups which show on the one hand a certain proportion of the male blood and on the other hand the same proportion of the female blood, it is possible to get back to the half blood basis. Thus, a male from group 7 mated with a female from group 8 results in the half bloods of group 10. Similarly a male from group 9 mated with a female from group 11 results in the half bloods of group 14.

It will be noted from the chart that in the line breeding it is possible by beginning with two individuals or two blood lines to keep the matings within these two lines, but to have a wide variation in the proportion of each in the stock. It is possible to breed back toward the male line to such an extent as to eliminate largely the female blood or to do the same thing toward the female line. When a group is secured with such a proportion or mixture of blood as to give outstanding results it is possible to continue the breeding so as to hold the same proportion of blood as long as desired or as long as there is no sign of detriment. This can be easily accomplished by selecting for a mating both males and females from the same group. Because each parent has the same proportion of blood, the offspring will likewise carry this same proportion. Thus, a male from group 3 mated with a female from group 3 results in offspring 5, which have one-half of the blood of male and female lines alike the same as the parent stock. In the same way a male from group 9 mated with a female from group 9 would result in offspring having 13-16ths of the blood of the male line.

While the chart shows that a male is selected from a certain group to mate with a female from another group to secure the results indicated, it must be understood that after the matings have progressed beyond the use of the original male or female, the reciprocal of such a mating may be used, with identical results so far as the proportions of the blood are concerned. Instead of mating a male from 7 with a female from 4 to produce group 9 having 13-16ths of the female blood, a male from 4 mated with a female from 7 would give offspring which would also have 13-16ths of the male blood.

The combination of 14 and 17 shows how new blood may be introduced if desired. A female from a new blood line mated with a male from 14 results in offspring 18, which have one-half of the blood of the new line, one-fourth of the blood of the original male line and one-fourth of the blood of the original female line. This new blood could be crossed in on individuals showing either more or less than one-half of the blood of either of the original lines, the effect being to increase the percentage of the blood of one of the original lines beyond one-fourth and approaching more nearly to one-half, with a corresponding decrease in the amount of the blood of the other original lines. A female from 17 mated with a male from 16 would result in offspring which are 32-64ths of the blood of 17, 21-64ths of the blood of 2, and 11-64ths of the blood of 1. A female from 17 mated with a male from 12 would reverse the proportion of the blood of 1 and 2, resulting in offspring which are 32-64ths of the blood of 17, 11-64ths of the blood of 2, and 21-64ths of the blood of 1.

Out breeding.—This term as used by poultrymen usually signifies the introduction of blood of some other strain or some other flock than that with which the breeder is working, but of the same variety. It is resorted to for the purpose of keeping up the vigor and vitality of the stock or in some cases in an attempt to improve the quality of the stock by the use of blood from stock which the breeder believes to be superior to his own. It must be kept in mind in making such a step that such out breeding has a tendency to destroy the uniformity of the flock in some respects and may have exactly the effect, although in lesser degree, that would follow crossing with a different variety or breed. The desirability of out breeding will depend upon the probability of an improvement from such a step and upon the necessity for introducing new blood to keep up vigor. Where the main purpose of the poultryman is the production of meat or eggs, and where he cares little for the uniformity of his flock from the show room point of view, it is common practice to out breed every year or two by the purchase and use of new males. As long as he uses males of the same variety as his flock and takes pains to select vigorous, healthy stock, this plan of breeding doubtless has advantages. It is a decided mistake, however, to use males of different varieties or breeds in introducing new blood, as it can only result in securing a product either in meat or eggs which lacks uniformity and consequently suffers in its market quality. Where a breeder has developed a strain of fowls of especially fine quality or where he has developed a high egg laying strain, he should hesitate a long time before he introduces new blood, and if he decides to do so, should use caution as to the method by which he does it.

Crossing.—By crossing is meant the mating of individuals of different breeds or varieties. This is sometimes resorted to by breeders for the purpose of strengthening or improving characteristics of color or type. An example of this is the use of Black Breasted Red Games on Golden Duckwing Games to strengthen the Golden Duckwing color. Crossing is also resorted to when it is desired to secure a new combination of characters in the development or establishment of new breeds or varieties. There is a belief that the mating of fowls of different breeds or varieties, and consequently those which are absolutely unrelated, tends to produce offspring which are unusually strong and vigorous, more so, in fact, than when crossing does not occur. There seems to be considerable foundation for such a belief. Crossing is therefore sometimes resorted to for the purpose of increasing vigor or vitality.

It must be clearly understood by anyone who contemplates crossing, that while the offspring directly resulting from the cross are often quite uniform in color, type and size, when these latter individuals are mated, their offspring usually vary very widely in color, type and size, and the variation may be all the way from individuals resembling one of the original parents of the cross in one or more particulars through many gradations and combinations to individuals resembling the other parent. Uniformity is very largely destroyed and can only be regained by the selection in successive generations of individuals embodying the desired characters. This is a long, slow process, and for this reason the crossing of varieties or breeds should be left in the hands of the expert breeder who has some definite and good reason for making the cross. Occasionally a cross is resorted to by poultrymen where experience has shown that the offspring will possess some desired quality in high degree, such as meat quality or good egg production, but where this is done it is necessary to make the same cross whenever young stock is desired. This is inconvenient, as it requires that stock of the varieties crossed be kept pure in order to have the individuals for making the cross.

Grading up.—This is a process of improvement of mongrel stock by the use in successive generations of pure or standard-bred males of the same variety. In each generation the females which most nearly resemble females of the variety of the male used should be selected and in turn bred back to another male of the same variety. Repeated practice of this sort leads to rapid improvement and in a few generations the stock can be brought to a point where for all practical purposes it is pure bred. In no case should males of different breeds or varieties be used in the different years, as such a step will destroy all the uniformity obtained and quickly degrade the flock again into one of mongrels. Nor should the grade males be used for breeding until the uniformity of the flock has been well fixed.

As a result of the first use of the pure-bred male, the resulting offspring carry one-half pure blood. In succeeding generations this becomes $\frac{3}{4}$, $\frac{7}{8}$, $\frac{15}{16}$, etc. It is easy to realize, therefore, why improvement is so rapid.

It must be clearly understood that grading up is not a practice which is recommended in preference to purchasing or keeping pure-bred stock. The latter course is by all means preferable. Grading up should only be practiced when it is impossible to make the necessary investment for pure-bred stock or where it is impossible to interest a poultrykeeper otherwise in the improvement of his stock. Grading up is most commonly practiced in improving farm flocks.

Introducing new blood.—In introducing new blood it is well not to use it indiscriminately on the flock until the breeder has determined how well it will blend or "nick" with the blood of his own birds. If a male is purchased he should be mated to three or four selected females and the quality of the offspring from this mating carefully observed. If the result is good, a male or males from this mating can be used on the flock. This will not only afford an opportunity to test out the fitness of the new blood for use, but by using the offspring of the trial mating will cut down the percentage of new blood introduced on the general flock, thereby lessening its disturbing influence, while at the same time securing in some measure such benefits as may accrue from the use of new blood. The purchased male after being thus tested for use on the breeder's flock, may, if the results of the trial mating seem to warrant, be used more widely the following season.

Many breeders prefer to purchase one or two females when introducing new blood. In this case they mate these females with a selected male from their own stock and if the result of this trial mating is good, they use the male offspring for the further introduction of the new blood.

Purchasing breeders.—In purchasing breeders it is by all means advisable to make a personal visit to the farm of the breeder from whom the purchase is to be made. This not only gives one a chance to pick out the particular kind of bird which he needs for his purpose, but also enables one to form a good estimate of the general quality and vigor of the stock as a result of looking over the whole flock. Such an inspection of the whole flock, together with a talk with the breeder, may often enable one to gain a valuable idea of the breeding tendencies of this particular flock, so that he can more intelligently mate the purchased birds with his own fowls. Where a visit to the breeder's flock is not possible, it is necessary to depend upon the judgment and ability of the seller to select the bird or birds which will do the purchaser the most good. Where such an order is given by mail, it is well to explain not only the kind of bird desired, but also to describe briefly but clearly the purpose of the mating which is to be made, and to give a good description of the purchaser's stock and its breeding tendencies, as this will often materially aid the breeder supplying the bird to select the one which will give best results.

It is suggested that the beginner in buying a trio or

pen from the breeder, purchase these already mated and request the breeder to explain why he puts those particular birds together.   It is also suggested that the breeder leg band and keep a record of any birds which he sells, and to whom sold, so that if he is called upon to furnish other birds to the same customer in the future he will be in better position to know just what birds to furnish for best results.

Often purchasers, particularly beginners, complain because birds purchased by them show some defects or weaknesses.   In this connection it must be borne in mind that no birds are perfect.   Weaknesses or things to be guarded against will be found in all birds to some extent, and one must not expect that birds purchased will be entirely free from them. In other words, one must not expect to receive a perfect bird and should not feel defrauded if some defects are shown. A bird which is not very bad in any one section, but which may show one or more slight defects which are not disqualifications,

Fig. 10. Barred Plymouth Rock female showing black feathers in the plumage. (Photograph from the Bureau of Animal Industry, United States Department of Agriculture.)

is a good bird and usually worth the money paid for it. Beginners also frequently complain that birds purchased by them show or develop a few off-colored feathers and think that this indicates impurity of blood. Such is not the case, for off-colored feathers are not uncommon in many varieties, as, for example, the black feathers in the Barred Plymouth Rock. Such off-colored feathers are not seen in the specimens exhibited at the shows, simply because they are pulled out, but their presence does not indicate impurity of blood. In chicks, also, before they get their mature plumage there may be foreign color of plumage shown, and this is often mistaken for evidence of impurity. With the growth of the mature plumage, however, this foreign color is usually completely lost. An example of this is the white which commonly occurs in the wings of chicks of black varieties, but which usually disappears from the adult plumage.

Standard or single mating. —By this is meant the use of birds of both sexes which approach as closely as possible the standard requirements for the variety in question. In some respects this is the simplest breeding to follow and is the method most commonly and most widely used, particularly in the case of solid colored birds.

Fig. 11. The same Barred Plymouth Rock female shown in Fig. 10, after the black feathers have been removed. (Photograph from the Bureau of Animal Industry, United States Department of Agriculture.)

Where this method is employed it is the expectation to secure birds of both sexes which are of good quality, or, in other words, which approach fairly closely to the standard. Many older and more experienced breeders are at present employing this method less and less exclusively, and are leaning toward the double mating system. In some cases, such as the solid colored birds, this double mating does not take the form of two matings which differ markedly in character, but rather in the development of two more or less distinct blood lines, one of which shows a greater tendency to produce males of good quality and the other to produce females of good quality.

Combination mating.—Frequently a special, or what may be termed a combination mating, is employed. In this mating one male is used with two different types of females, from one of which standard male offspring are more particularly expected, and from the other standard female offspring.

Double mating.—Double mating consists of the use of two distinct matings or sets of matings; one for the purpose of producing a greater percentage of standard males, and the other for the purpose of producing a greater percentage of standard females. The first of these matings is therefore known as the cockerel mating, and the females from it will usually prove unsuitable for exhibition, but are useful for continuing this line of breeding, and are known as cockerel-bred females. The second of these matings is known as the pullet mating, and the males produced from it will usually prove unsuitable for exhibition, but are useful for continuing this line of breeding, and are known as pullet-bred males. It will, therefore, be seen that the practice of double mating necessitates the carrying of two distinct lines of blood, the pullet line and the cockerel line. Often the matings are quite distinct in character, as in the case of the Barred Plymouth Rock, where each mating is made and designed to offset the natural tendency of the male and female offspring to diverge in their color markings. However, as mentioned under Standard or Single mating, double mating sometimes resolves itself simply into the keeping of two distinct lines of blood which have proven especially suited, one for the production of males and one for the production of females.

A fixed plan of breeding.—For success in breeding it is necessary first of all to have a definite ideal in mind. In case this is some utility quality, such as increased egg production or improved table birds, the breeder must have clearly in mind just what he wishes to accomplish. In case it is the production of standard or show specimens, he must carefully study the description of his variety, as given in the Standard of Perfection, and become so familiar with these requirements that he has the ideal clearly and definitely fixed in his mind. Having a definite ideal to breed to, he must then have a fixed plan of procedure, and as this is important, he must in its main features adhere to it. He will fail to make progress of a substantial character if he continuously changes his plan of breeding, and his success, instead of consisting in producing a good percentage of birds of quality, will be, if any, of a spasmodic and accidental kind that produces only an occasional good individual. A measure of the success of the breeding is a comparison of the young stock with the parent stock. Unless the young stock shows on the whole an improvement over the parent stock, no progress is being made. If such a lack of improvement continues it is safe to assume that the plan of breeding is wrong, and a change of method should then be made.

Foreseeing bad tendencies.—The successful breeder must be foresighted. He must be quick to see any bad tendencies creeping into the flock and must take steps in his matings to correct them. The quicker he notes such tendencies, the easier will be his task of eradicating them.

If he does not recognize the danger until it shows strongly in the flock, he will have a serious time eradicating it. This again emphasizes the necessity for the breeder to be continuously studying his flock in the most minute manner. He must in effect be thinking and planning ahead of his actual matings. For example, a breeder of White Leghorns may notice a troublesome tendency toward white in face creeping into his flock. He must then immediately become more rigid in his selection against white in face and use no birds showing it if it can possibly be avoided. In any case, it will be a wise precaution for him to make a special mating of birds strong in face, so that he will have available a quantity of young stock, especially strong in this particular, from which he can select breeders to use in his future matings and so check this troublesome tendency.

Establishment of blood lines.—The breeder must aim to establish his own blood lines. To do so, he must concentrate on the lines which show the most promise in his breeding. When blood lines are fairly well established he can rely much more certainly on the results to be obtained from the matings than if blood of different lines is continuously being introduced. Blood from some other line may utterly fail to blend or "nick" well with his own lines and the result of such a mixing of blood lines may often have the same upsetting results, though within narrower limits, of course, that is obtained in crossing varieties or breeds. When it is desired to seek improvement by introducing new blood lines, this should be done in such a way as to upset the breeder's own lines as little as possible. See section on introducing new blood.

Offsetting the weak points of one sex with strength in the other sex.—This is a pretty generally recognized practice of breeding and is well founded. Having determined to use certain females or a certain male in the mating, the selection of the individuals of the opposite sex should embody in so far as possible, strength in those sections where weakness is shown in their mates. Failure to do so may, and usually does, result in the intensification of the weakness, and may finally result in the presence of a weakness or defect which is common and firmly fixed throughout the flock. Where it is not possible to make such a compensating selection in a mating, other matings should be made, which will be strong where the first mating is weak, so that the resulting offspring can be used in succeeding matings to correct or offset the weakness.

Grouping the birds for selection of the matings.—When selecting birds for a mating it will be found a material help if they can be brought together so that they can be easily and carefully compared. If a few exhibition coops are available the birds can be placed in these and the comparison easily made. The females or the males can then be brought side by side, studied and handled and a more intelligent selection made than if the birds are separated in different pens and the memory is relied upon to select those best suited for the purpose.

Breeding birds with defects.—The advisability of breeding birds with defects depends upon the seriousness of the defect and upon the freedom of the stock as a whole or its tendency toward the same defect. If the bird possessing the defect is an outstanding individual in other respects, it is often advisable to breed it. It is better to discard than to use it, however, if the flock shows a tendency toward this defect. In any case, when such a bird is used, the opposite sex of the mating should be selected with especial care to offset the defect. Many defects are due to accident or certain forms of deformity. Red spots in white ear lobes, crooked breast bones due to too early roosting, or the occasional crooked or roach back that is thrown by normal individuals need cause the breeder little or no concern.

Breeding birds with disqualifications.—In general, it is unwise to breed birds showing disqualifications. In the first place, a disqualification if reproduced renders the birds unfit either for sale or for exhibition, and in the second place, disqualifications are likely to be defects which are more or less common and troublesome to the breed or variety, and therefore quite likely to be reproduced. Sometimes a disqualified bird of such extraordinary excellence,

except for the disqualification, is obtained that it is desirable to breed it. The individual judgment of the breeder must decide whether this is wise, and, as in the case of birds showing defects, the advisability will depend largely upon the commonness of the occurrence of that particular disqualification in his stock. Disqualifications such as stubs

Fig. 12. Single comb showing side sprig. (Photograph from the Bureau of Animal Industry, United States Department of Agriculture.)

and side sprigs are common and strongly inherited in some breeds, while in others they are rare and less strongly inherited. For example, it is dangerous to breed a Rhode Island Red showing stubs or side sprigs, less dangerous to breed a Barred Plymouth Rock under the same circumstances, and still less dangerous to use such a Leghorn. In any case, only super excellence of a disqualified individual justifies its use as a breeder.

Defects likely to occur in common color matings.—In varieties of the same color but of different breeds there are certain well-defined defects which are more or less common to all of them and which must therefore be uniformly guarded against in the mating. The defects mentioned in connection with the common color matings given below are likely to appear in that colored variety of any breed, but may occur more commonly in some than in others. They are grouped under these headings for the information of anyone who is particularly interested in a general color mating.

White matings.—Black or dark ticking; creaminess in quill or undercolor; brassiness; black or foreign color in the quills; solid black in wings or tail; black feathers in any section; red, buff or salmon in the plumage; gray plumage.

Black matings.—Gray in the hackle and saddle of males; red, straw or silver in the hackle, back, wing bows or saddle of males; white, gray or light undercolor in the hackle, back and saddle of males, especially in hackle; white in sickles of males; bronze tinge on the shoulders and especially on the tails of males; brownish or dull black surface color in females; white or gray in wings; purple barring; white or gray at root of tail; dull black surface color lacking the green sheen; frosted or white tips to the wing feathers; white in flights.

Buff matings.—White edging to the sickle feathers of males; unevenness of color of the hackle, back, wing bows and saddle of males; too heavy color on the shoulders and back of males; too light or white undercolor at base of hackle, at base of tail or in the saddle of males; wing bow darker than breast, especially in males; tendency for the saddle and tail coverts to be laced with white; shafting, mealiness and patchiness in females; too light color in females; white, black, a peppering of either or smokiness in the wing flights or in the main tail feathers; white in under-color; white or very light buff quills close to the body; feathers tending to be laced.

Columbian matings.—No lacing on the back at the base of the tail in males; brassy surface color in males; black ticking on the throat, breast and fluff of males; purple bar-ring in the black of male's tail and hackle; brown cast to the primary wing feathers of males; black or black ticking in the surface of the backs of females; too heavy lacing in the tail coverts of females; indistinct or smutty lacing in the hackle and tail coverts of females; white spots in the pri-maries of females; black in the surface of the feathers at the sides of the fluff and of the body feathers just in front of the fluff; too light or faded appearing markings; the white lacing not extending clear around the ends of the hackle feathers; too light or pure white under color; main tail and sickle feathers not black clear to the skin; gray in the wings.

Golden mating.—Smutty wing bars and shoulders; frosting, especially on the breast; mossiness both in pullets and in hens that have molted mossy; the lacing of the hackle not extending around the ends of the feathers; purple barring in the black sections of the males; too heavy lacing; black peppering or mossiness in the ground color; black peppering in the wings; light or white in the black of the flight and main tail feathers; general color of the plumage too light; uneven lacing; lacing weak or lacking under the throat and on the head.

Silver matings.—Smutty wing bars and shoulders; too narrow or too heavy lacing; frosting, especially on the breast; mossiness, both in pullets and in hens that have molted mossy; the lacing of the hackle not extending around the ends of the feathers; black peppering or mossiness in the ground color; black peppering in the wings; general color of the plumage too light; uneven lacing; lacing weak or lacking under the throat and on the head.

Partridge matings.—The black striping running through the ends of the hackle and saddle feathers of the males so that there is no red edging showing clear around the ends of the feathers; white in the tail and wings, especially in males; too light colored females with lemon hackles; stippling in the tail coverts of females; too dark a red in males; lemon hackle in males; yellow shafts to the hackle and saddle feathers of males; purple and purple barring in the black sections of males; white or light in the under color of the hackle, back and saddle of males; metallic sheen in females; too dark penciling, that is, the black pencilings wider than the red of the feathers; too fine or too narrow penciling; stippling, barring or broken penciling in the back and fluff; dark legs, particularly in females.

Silver Penciled matings.—Red on wings of males; solid white in flight and main tail feathers of males, other than the white edging of the primaries and secondaries called for by the standard; brassy backs of males; inclination to brownish cast instead of gray in females, which is apt to increase with age; gray in flight feathers; brown on the shoulders of males; purple barring in the black sections of males; indistinct penciling in females; poor penciling of the throat and breast of females; poor penciling of the lower back and tail of females; the hackle and saddle striping running clear through the ends of the feathers.

Mottled matings.—Birds too light in color; too large white tips of the feathers; indistinct white tips showing some black or gray; too many white feathers in the wing bows, both of males and females; red or brass in the hackle, back, wing bows and saddle of males; too much white in the wing flights and secondaries and in the main tail feathers; ashy color in the white tips; purple barring; white feathers in the bodies of females.

Use of tested breeders.—Often it happens that some individual, which may or may not be in itself an exceptional individual, will prove to be unusually prepotent and consequently have an unusual value as a breeder. This can only be determined as the result of matings made, but when such an individual is found it should, of course, be used as long as it will breed to advantage. Sometimes a particular mating will be found to give unusually good results, and when this is the case the same mating should be kept together in successive years as long as possible.

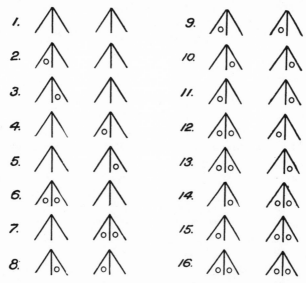

Fig. 13. The combination of marks possible to use in toe-punching chicks. (From the Bureau of Animal Industry, United States Department of Agriculture.)

Pedigreeing.—Young stock should be pedigreed. This means that a record be kept of their parentage so that it is possible to know the parents of any desirable individuals. Only by pedigree breeding can the breeder know what the different matings or the different individuals are producing, and consequently their breeding value. The most common form of pedigreeing is simply to keep a pen record. This is most commonly done by toe punching the chicks from each mating with a distinctive mark. Chick leg bands may be used for this purpose if desired, but toe punching is simple and answers the purpose very well. There is a tendency for some of the toe punches to grow up, and to make sure that this does not occur and thereby confuse the record, the chicks should be gone over carefully when about two weeks old, and any holes which show a tendency to grow up punched out anew. This second punch, if necessary, is almost certain to be permanent. See Fig. 13. These pen pedigrees tell the sire of the chicks, but do not indicate which hen they are out of.

Where pedigree records are desired which will give the parentage on both the male and female side, it is necessary to resort to a more complicated system of records, except where the matings are few and the individual hens of all the matings involved do not exceed 16 in number. For any number of hens up to 16 a different toe punch can be given the chicks from each hen.

Fig. 14. Eggs of individual hens placed in pedigree basket and bag ready for hatching. (Photograph from the Bureau of Animal Industry, United States Department of Agriculture.)

Trapnesting is the first requirement of complete pedigreeing in order to be able to identify the eggs from each hen. These eggs should be marked, preferably on the large end, with the number of the hen, and if desired, with the pen number as well. Sometimes the system is used of writing the pen number, then drawing a line and beneath writing the hen number, e. g., 9/36, 9 being the pen number and 36 the hen number. If the eggs are to be incubated under hens, enough must be saved from an individual hen for a sitting, or eggs from two or more hens of different breeds may be used to make up a sitting. If they are to be hatched in incubators, eggs from the different hens are placed in the machine together. On the eighteenth day the eggs from each hen are sorted out and placed in a pedigree bag or pedigree trays for hatching. Pedigree trays, consisting of wire-enclosed compartments to be set on the ordinary incubator trays, can be purchased. Suitable pedigree bags can be made easily and cheaply of mosquito bar or bobbinet of any size desired and consisting of one, two or four compartments. See Fig. 14. The eggs are placed in these bags, which are tied tightly shut so that the chicks cannot escape and become mixed. A fairly liberal allowance of room should be made in these bags, for if the eggs are crowded they will not hatch so well and the chicks are more likely to be crippled.

Fig. 15.   Chick with duplicate pedigree leg bands. (Photograph from the Bureau of Animal Industry, United States Department of Agriculture.)

As soon as the hatch is complete the chicks are removed from the bags and each one banded with an open numbered chick band such as can be purchased of any poultry supply house. A note is made of the band number and of the parentage of the chick bearing this band. The band must be pressed around the legs of the chick closely enough so that it will not slip off over the foot. See Fig. 15. As the chick grows, the band must be opened from time to time to allow for the increased size of the leg. The growth of the leg will not force open the band, and failure to attend to the bands will cause the leg to grow around and beyond the band and may cause a bad sore or even the loss of the chick. When the chick is two to six weeks old the band should be removed from the leg and placed in the w i n g, where it can be worn until maturity. Fig. 16. A small knife blade should be run through the thin skin on the upper side of the wing which extends from the shoulder j o i n t to the outer joint of the wing,

Fig. 15. Position of pedigree band after it has been transferred from the leg to the wing. (Photograph from the Bureau of Animal Industry. United States Department of Agriculture.)

and the band passed through this hole and pressed together. By holding the spread wing up to the light a spot can easily be selected about one-quarter inch from the outer edge of the skin where no blood vessels of any size are visible, and little or no bleeding will result. Fig. 17. The ends of the band should be brought together or lapped slightly and pressed close together, but the whole band should be left rounded and not to be pressed tightly together so as to pinch the skin, as this may cause irritation, swelling and pain, or may cause a sore which may lead to the band sloughing off and becoming lost. These bands stay on the legs and in the wings of the chicks remarkably well and only a small percentage are lost. If desired, two bands bearing duplicate numbers can be used, one on each leg, and later transferred to each wing. This will reduce the numbers of chicks with lost pedigree records to a minimum. Pedigree breeding in-

Fig. 17. Place where pedigree band is inserted through the wing. (Photograph from the Bureau of Animal Industry, United States Department of Agriculture.)

volves a considerable amount of extra work and detail record keeping. It is, however, the only means by which the breeder is able to determine with certainty which birds and matings are producing results and which are failures. For the breeder who is striving for improvement and who wants to know exactly what he is doing, whether in breeding for standard requirements, for utility, or for both, it is well worth while.

Record of matings.—A record of the matings made each year should be kept. Each pen or mating should be given a number, and under this should be listed the females and the male used in the mating. The value of this record lies in the

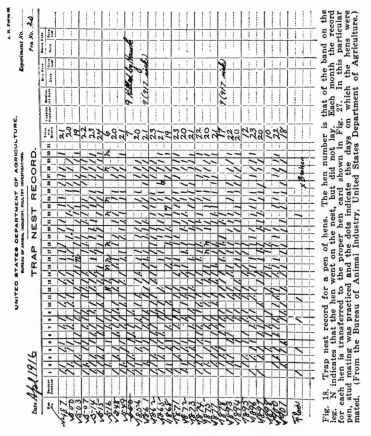

Fig. 18. Trap nest record for a pen of hens. The hen number is that of the band on the leg. N indicates that the hen went on the nest, but did not lay. Each month the record for each hen is transferred to the proper hen card shown in Fig. 27. In this particular pen, stud mating was practiced and the dots indicate the days on which the hens were mated. (From the Bureau of Animal Industry, United States Department of Agriculture.)

fact that if a mating gives unusually good results, so that it is desired to repeat it the following year, this can be done without any uncertainty as to the individuals involved.

Description of matings.—A description of the matings and of the individuals used in the matings should be made and filed for reference. Such a description may prove very valuable, as it enables the breeder to refresh his memory as to the exact character and peculiarities of any individuals which have given good results even if they have died, and may then yield him valuable breeding knowledge which he would otherwise overlook. Such a description, when made of pullets and cockerels, will give the breeder a knowledge of the bird in pullet or cockerel condition, and may enable him to know in later years whether certain defects are due to age, and will guide him, therefore, as to whether it is wise or unwise to use them in any particular mating which he has in mind.

When chicks are pedigreed and a record kept of the matings, it is well to go over the young stock critically in the fall before the old stock is culled. By doing this, a good idea will be gained as to which of the breeding stock has given results in superior young stock and which have failed to do so. With this knowledge, the old stock of breeders can be much more intelligently culled.

Stud mating.—When it is desired to breed the females of a flock to two or more males, and where separate pens are not available for this purpose, stud mating is sometimes resorted to. Stud mating refers to the practice of keeping the males used in separate quarters or coops and bringing to them the females with which they are to be mated as they lay. Consequently trapnesting is a necessary part of stud mating. If a hen is mated after each second egg she lays, the fertility will be equally as good and may be better than if the male ran with the flock. See Fig. 18. In fact, a mating after every third egg laid seems to give satisfactory results. The hen need be left with the male only long enough for copulation to take place, but since it is usually not expedient to wait around and assure one's self that this has occurred, leaving the hen with the male for a half hour is adequate. Stud mating requires more work than ordinary mating. Its advantage lies in the ability to mate the different females of a flock with different males, while allowing the hens to remain together and be cared for as a single flock. It also insures better fertility where it is desired to breed only a few females from a large flock than would be obtained if a single male were allowed to run with the hens. Where a flock is too large to be taken care of by a single male, and where two males cannot be used at one time or alternated because definite pedigrees are desired of the chicks, it is also useful.

Alternating males.—With flocks too large to give good fertility with a single male, two or more males are often used. They can be allowed to run with the flock at one time, but in this case there is a tendency for the strongest male to dominate the others and to do most if not all the breeding. Better results are obtained where the different males used are alternated. This is accomplished by allowing one of the males to run with the flock in the morning and the other in the afternoon, or else upon successive days. Many poultrymen use this system even with smaller flocks, believing that they get better results in fertility by this practice. The disadvantage of such a system is that it is obviously impossible to determine which male is the sire of any of the offspring.

Age of breeders.—Only fully matured birds should be used as breeders. Yearlings or older hens make better breeders than pullets because they lay larger eggs and also because they have had a period of rest during the molt preceding the breeding season, while the pullets, if they have been laying all the fall and winter, are more or less exhausted and consequently in less satisfactory breeding condition.

Well-matured cockerels, because more active, give better fertility than older males. To secure best results in breeding it is well to use cockerels with hens and comparatively

young, vigorous cocks with pullets.   Sometimes a young, undeveloped cockerel is of such excellence that it is desirable
to breed him.   By placing such a cockerel with a couple of
hens for some time before eggs are to be saved, he will be
developed faster than if kept away from females.   The
number of females mated to an undeveloped cockerel should
be small, never exceeding three or four.   Where cockerels
are run together and separated from the females, it is usually desirable to pick out the most promising and to separate
them from the others, either by placing with hens or by returning them to the pullet range so as to give them a good
chance to develop to the best advantage.

Size of mating.—The size of mating will depend upon the
breed, the conditions under which the birds are kept and
the condition and vigor of the birds themselves.   One male
of one of the lighter breeds, such as the Leghorn, will usually give good results with about 15 females; a male of the

Fig. 19.   A good range means well-grown young stock.   (Photograph from
the Bureau of Animal Industry, United States Department of Agriculture.)

general purpose breeds, such as the Plymouth Rock, with
10 to 12 females, and a male of the heaviest breeds, such as
the Brahma, with 8 to 10 females.   Where the male is
especially strong, vigorous and active, and where the fowls

are given free range, the number can often be materially increased with success. A Plymouth Rock male on a flock of 20 females and a Leghorn on a flock of 30 to 40 females on free range have frequently given splendid fertility.

Breeding condition.—Fowls to be in good breeding condition should be well fed and be in good flesh, but not excessively fat. Thinness or overfatness are both detrimental alike to egg production and to fertility.

Early hatching.—This is an important element in the successful rearing of young stock of good quality. Not only do the chicks hatched early live and grow better, but early hatching is essential if young stock of the desired size and finish are desired for the fall and early winter shows. The hatching should be done in February, March and April and should be completed by the first or at latest the middle of May. Early hatching also has a most beneficial effect upon the egg production, for it is only the early hatched pullets which mature in the fall in time to start laying before the cold weather sets in, and are in consequence the profitable winter layers.

For the late winter shows, later hatching is an advantage in order to have the young stock at the very height of condition at that time. It is also true that late hatched birds may sometimes appear to be better colored, due to the fact that their mature plumage was acquired later and has not been subjected to much intense sunshine, which may cause fading.

Free range vs. confinement.—There is absolutely no question as to whether or not free range is preferable for breeding stock. Not only will the fowls so allowed to range be in better health and condition, but the eggs will show better fertility and will be larger and capable of hatching stronger chicks. If it is not possible to give a breeding flock free range, they should be given as roomy a run as possible and should be regularly and plentifully supplied with green feed, such as kale, cabbage, beets, clover, alfalfa or sprouted oats,

Fig. 20.   A wire-covered frame which allows the smaller chicks access to the feed while excluding the larger chicks. (Photograph from the Bureau of Animal Industry, United States Department of Agriculture.)

Good management.—Good management and good breeding are so closely dependent that without both the best results are impossible of attainment. The young stock must be carefully and plentifully fed, well grown and kept free from lice and mites. If this is not done, birds of the desired size and of the desired finish and condition will not be obtained. Insufficient feeding and poor or irregular growth are frequently responsible for certain defects, not only of feather growth, but also of color.

Culling.—The young stock must be severely and courageously culled. A successful culler and a successful

Fig. 21. A supply of dry mash where the young stock can help themselves promotes rapid, even growth. (Photograph from the Bureau of Animal Industry, United States Department of Agriculture.)

breeder are almost synonymous terms. The breeder must recognize what he wants in his young stock and must have the courage of his convictions to cull out those birds which do not meet his requirements, and which, while admirable in some respects, have serious defects or disqualifications which are bound to prove troublesome if the bird is retained and bred. Continuous culling throughout the season will gradually eliminate the young stock of poorer quality and will make room and give the better specimens a better chance to grow and develop.

The experienced poultryman will also cull his entire flock the year round. Whenever he sees a bird which is going to pieces or which is badly out of condition or sick, he will cull it, and thereby not only get rid of unprofitable birds, but reduce to a minimum the chance for disease to spread through the flock.   In addition, most breeders cull out those birds which they no longer wish to carry as breeders after they have stopped laying in the summer or fall.

Trimming heavy feathered fowls.—Where fowls are extremely heavily feathered, like the Cochins and some Wyandottes, the profusion of feathers about the vent and tail frequently interferes with successful copulation, with the result that the fertility of the eggs is poor. To remedy this condition, cut or trim the feathers away about the vent and also a little off the main tail feathers of both sexes just before the breeding season. Often this will result in much better fertility.

Fig. 22. Single Comb White Leghorn cockerel out of hen 408, who is shown with her record in Fig. 24. The sire of the cockerel was also out of a hen which laid 202 eggs in her pullet year. (Photograph from the Bureau of Animal Industry, United States Department of Agriculture.)

Combination of utility and quality.— A great amount of discussion has arisen over the question of utility vs. standard bred poultry. There seems to exist a fairly wide opinion that if a breeder is working for utility he cannot have birds of good quality, while if he is working for quality his birds must of necessity be poor utility fowls. There is no good foundation for such an opinion. Many hens of excellent quality have been decidedly superior egg producers, while the standard shape requirements for the different breeds insure good table type. Admittedly it is harder to secure fowls showing a combination of excellent quality and excellent producing ability than to secure either one alone, just as it is always harder to select successfully for two characters than for one, but that such a combination does frequently exist and can be obtained there can be no doubt. It is being done by breeders and the authors themselves have succeeded in securing this result in their breeding work at the Government Poultry Farm, Beltsville, Maryland. It is equally true that for the breeder who can secure such a combination of show quality and high egg production there is a broader field than for breeders who specialize on one line alone, and just so surely as it requires greater pains and superior ability to secure this result, just so surely will the demand for his stock and his financial reward be greater.

CHAPTER III

BREEDING FOR INCREASED EGG PRODUCTION

Importance of vigor and health.—Egg production, and in particular heavy egg production, is a severe strain upon a hen or pullet. Any individual, therefore, which is lacking in health, strength or vigor is unable to stand up under this strain and is in consequence unable or extremely unlikely to make big records. For this reason it is especially important that the breeding stock be very strong, healthy and vigorous if their offspring are to show these same qualities and be so constituted that they can bear the strain of heavy egg production, provided they have the inherent ability. Therefore, in selecting breeding stock with the idea of improving the average egg production of a flock, it is of primary importance to be sure that the breeding stock shows strong constitutional vigor and that they are in the best of health. Strength, constitutional vigor and health are usually indicated by certain characteristics of fowls. Fowls having these charcteristics should be active, alert and spirited; the males in particular should have a fearless appearance, the eye should be bright and fairly prominent and the birds should stand strongly on their legs. They should have bone of good size and strength for their breed and should show by the bright red color of head parts and by the smooth, well-kept condition of their plumage that they are in the best of health. Avoid birds which are listless and inactive, are weak on their legs, have dull eyes, long or snaky heads and sunken eyes.

In breeding for improvement in egg production it frequently happens that in order to get ancestry on both sides which has high production back of it, the fowls are rather closely inbred. While it may at times be necessary or desirable to make such inbred matings, it is best not to continue this practice too long without introduction of blood less closely related. It seem to be fairly well established that the crossing of two breeds or varieties tends to result in improving the vigor of the offspring, and this same result is usually apparent also when fowls of the same breed or variety, but of distinct and unrelated strains, are mated together.

Professor James Dryden of the Oregon experiment station, in his work in breeding for increased egg production, has found that the increased vigor obtained by using unrelated blood adds very materially to the result in egg production simply from this cause alone. In his work, however, he has not been contented to stop with the improvement so obtained, but

Fig. 23. Single Comb White Leghorn cockerel out of hen 2033. This hen laid 230 eggs from November 1 to November 1 of her pullet year, and 260 eggs before she stopped laying. Eight of her sisters and half-sisters produced as follows in their pullet year: 180, 187, 195, 196, 198, 208, 235 and 240 eggs. The sire of this cockerel is a son of hen 514, which laid 213 eggs in her pullet year and 536 eggs in three years. The grandsire on the sire's side is a son of hen 408, who is shown with her record in Fig. 24. (Photograph from the Bureau of Animal Industry, United States Department of Agriculture.)

has followed up this increased vigor by careful selection of individuals according to their performance and breeding qualities, and has been able to improve egg production to a considerably greater degree than resulted from the increased vigor alone. While in some of his work he has crossed breeds with good results so far as the egg production of the offspring is concerned, he also states that it is unnecessary to cross breeds, as he can secure the same results by crossing unrelated strains within a breed or variety and by the proper selection of individuals in succeeding matings.

How high egg producing ability is inherited.—There has been a considerable difference of opinion as to how high egg production is inherited. It was the commonly held belief that this quality of high production might be inherited either from the male or female parent, or from both. Later, however, Dr. Pearl, as the result of his work at the Maine experiment station, concluded that one of the factors which are responsible for high egg production is inherited from generation to generation in accordance with Mendel's law, and he further concluded that this particular factor was sex limited, or, in other words, that it could not be inherited directly by pullets from their dams, but could be inherited by them from their sire only, who in turn might get it either from his dam or his sire. Other investigators working along the same lines do not agree with Dr. Pearl's conclusions, feeling that their results show that to some extent, at any rate, high egg production can be inherited directly from the dam as well as from the sire. It must be remembered also that even if Dr. Pearl's conclusions are correct, and if it is necessary for pullets to inherit the ability to be high producers from their sire and not directly from their dam, it does not follow that the egg producing ability of hens should be disregarded in breeding, since cockerels may inherit this quality from their dams as well as their sires, and it is only the cockerels which inherit this quality from both sides that are of the highest prepotency in transmitting it in turn to their daughters.

In general, the problem of breeding for improved egg production is a very complex and difficult one, but it may be reduced in its simplest terms to the proposition of breeding from the higher producers in the flock and discarding as breeders the lower producers. It does not appear, in the opinion of those who have had experience along this line, to be necessarily good policy to breed from the very highest producers or most outstanding individuals in this particular, but rather to breed from individuals whose production is high but does not run to the extreme. It is recognized to be more or less of a general truth that individuals which are well above the average in any particular quality, but are not the outstanding extremes, are more likely to transmit their good qualities to their offspring than are the extreme individuals.

In attacking the problem of what may be done in the way of breeding for improved egg production, it is necessary to consider the various conditions under which poultry are kept and to consider from this what may be the practical steps to take under existing conditions.

Breeding from an untrapped flock.—Flocks of poultry kept for egg production are in the main cared for without any detailed records being kept. Usually there will be no records available of the egg production of the different pens and in many cases even of the flocks as a whole. Under these conditions, of course, the hens are not trapnested and no information is available as to the individual breeding of any of the birds, either males or females. All that can be done, therefore, is to select from the general flock as intelligently as possible the best individuals to be used as breeders.

First of all, in this case the birds must be selected which show every evidence of being strong and healthy and which are possessed of good constitutional vigor so far as this can be determined from external indications. It is also necessary that the birds have a shape of body that is of sufficient length, breadth and depth so that there is plenty of room for a good development, both of the digestive organs and the egg producing organs, or, in other words, that have good capacity. Fortunately also, there are certain external characters which form fairly reliable guides in picking out the better producers from a flock. The best time of the year to make such a selection is in the late summer or during the fall, October or November usually being most satisfactory. It will be found that birds which at this time of the year are still laying are the birds which in the main have been the better producers of the flock for the entire year. This applies most satisfactorily to hens at the end of their first or pullet laying year.

Selecting good layers from the untrapped flock.—Indications of good laying ability or of a good yearly egg pro-duction are those things which show that a hen is still laying at this time of year: First is the failure to molt until late in the fall. Hens or pullets which do not molt until that time are, other things being equal, still producing eggs, and will usually prove to have been the best producers throughout the entire year. The condition of the

Fig. 24. Single Comb White Leghorn hen No. 408. This hen laid 214 eggs in her pullet year and 654 eggs in 4 years. (Photograph from the Bureau of Animal Industry, United States Department of Agriculture.)

molt is one of the outstanding characteristics in determining egg producing ability. With pullets it may be said in the vast majority of cases that those which molt before the first

of August are the poorest layers and should be culled from
the flock. Those which begin their molt between the first of
August and September 1 should likewise be culled, but are
ordinarily not as poor layers as those molting before this
period. Pullets which begin to molt during the month of
September, other things being equal, should be retained as
fairly good layers, while those which do not molt until after
October 1 may be considered as the best producers in
the flock.

While this distinction is not nearly as clear with hens, at
the same time it applies in a general way, except that hens
usually molt earlier than pullets. Comparatively few hens
molt as late in the fall as do the best laying pullets.

The second is the condition of the comb. When a hen is
laying or about to lay, her comb is large, full of blood and
bright red in color. When not laying, the comb is small and
shrunken, pale or dull in color, comparatively hard and
covered with whitish scales. A dark or bluish color usually
indicates sickness. Changes in the wattles and ear lobes are
similar to those of the comb, but not so marked.

The third is the spread of the pelvic bones. As a hen
stops laying, the pelvic bones, that is, the two bones which
can be felt as points on either side of the vent, approach
closer together than they are when she is in full lay. In all
probability if the spread of these two pelvic bones is two
fingers or less she is not laying, while if the spread is greater
than this she is probably laying. Of course, in measuring
this spread, the relative size of the hens of different breeds
with the corresponding natural difference in the spread must
be kept in mind.

The fourth is the distance from pelvic bones to keel. As
the hen stops laying there is also a decided tendency for the
distance from the pelvic bones to the rear end of the keel to
decrease, due to the fact that both egg organs and intestines
are smaller and require less room. A spread between these
two points which measures three or more fingers in the

smaller breeds and four or more fingers in the larger breeds usually indicates that the hen is in a laying condition, while a spread of less than this indicates that she is not in a laying condition.

The fifth is the flexibility of abdomen. When the hen is laying, the greater size of the abdomen, which is associated with a greater spread from pelvic bones to keel, and the lessened tendency to accumulate fat in the abdomen results in a softer and more flexbile abdomen than is the case with a hen which is not in a laying condition. The abdomen of a laying hen suggests when handled the texture of a partly milked out udder of a cow, while that of the hen which has ceased laying feels harder, more compact and less flexible.

The sixth is the appearance of the vent. The vent of a hen which is laying heavily is large, expanded and moist, while that of a hen not laying is comparatively small, hard, puckered and dry.

Seventh, shank and skin color. In breeds which have yellow skin and yellow legs there is a decided tendency for this color to be lost or to fade out as the hen lays. The rapidity and degree to which the yellow color is lost depends to a considerable extent upon the heaviness of laying. However, hens which show strong or medium yellow shank color in the fall are almost certainly poor layers, while those which show a bleached or white color of leg may be good layers, athough not necessarily so. Hens which are kept on grass range do not lose the shank color as quickly or completely as those kept in yards, while some soils tend to bleach the color of the legs and may cause even poor layers to have white legs. Moreover, a sick hen, or one in poor condition, may show pale shanks.

As is the case with the shanks, so the beak loses its yellow color as the laying progresses. The beak color, however, is lost more quickly than the shank color and is also regained more quickly after the molt. The shank color is therefore a better indication of egg production over a long period. In the case of yellow skinned breeds, the skin immediately around the vent loses its yellow color very quickly with laying and regains it quickly after laying ceases. A white or pink vent color, therefore, generally indicates that the hen is laying, while a yellow vent color indicates that the bird is not laying. In forming an opinion in the fall as to whether or not a hen is laying, and consequently in judging whether or not she has been a good layer for the entire year, it is best not to depend upon any one of these indications alone, but to consider all of them to see whether a majority does not agree in indicating laying condition and good production. It is necessary also to guard against the selection of occasional hens which may be laying at the time but have not laid heavily. The condition of the plumage will indicate a hen which has molted and resumed laying, while the shank color will usually serve to detect a poor layer. By selection on this basis not all good producers will be included, nor will all medium producers be thrown out, but those selected will nearly all be good producers and practically no poor producers will be included.

Where no records are kept, therefore, of individual performance or even of flock performance, it is possible by means of the principles indicated above to select from the flock the strongest, healthiest and most vigorous females with a body so shaped as to give them plenty of room for the vital processes necessary to heavy egg production, which are within reasonable limits of certainty the best layers in the entire flock. These females should be separated from the rest of the flock and mated to a good, strong, vigorous male, out of a hen, if possible, which is known to be a good producer. The continued process of the careful selection of breeders along these lines, together with the use of males from the selected flock, will result in breeding from the higher producers in the flock, and should give an improvement in the egg production.

Testing the offspring.—Where more than one pen of

breeders is used, it is very desirable to toe punch or otherwise mark the chicks from the different matings so that it will be possible to identify the mating from which they came. As the pullets from these various matings are put in the laying pens in the fall it is well to separate them so that the pullets from the different matings are placed by themselves. By the simple process of keeping a record of the egg production of the pens it is then possible to tell which of the matings is giving best results so far as the egg production of the pullets from them is concerned. Needless to say, if it is discovered that the results from any particular mating are outstandingly better than those from any other mating, this same mating should be put together the following year and for as many years as the birds are in vigorous breeding condition and the greatest possible number of chicks produced from them. Such other breeding stock as it is necessary to use should also be from the original pen or pens from which exceptionally good results were obtained.

Breeding from pen pedigreed flocks.—Where the fowls are kept in separate pens so that different pen records of egg production are available, the females for the matings in breeding for improved egg production should be selected on the same basis as stated above, since there is no surer way of determining which are the highest producers, as individual records as the result of the use of trapnests are not available.

Where pen records are kept, and particularly where the breeding pens are small enough so that only a single male is mated to the hens in the pen, it becomes possible by toe marking or otherwise to identify the chicks in so far as the identity of their sire is concerned. This makes it possible to secure a good line on the breeding ability of the different males used in the breeding pens by proceeding as indicated above, and to pen the offspring from the different pens separately and to keep a pen record of the different lots so as to test the breeding value of the sires.

Needless to say, any male which may be discovered to be a particularly prepotent individual with respect to the characteristic of high egg production should be preserved and mated as long as he is in good breeding condition and his blood should be spread through the flock as widely as possible. This makes it necessary to hold the males used in breeding until their daughters have been tested for egg laying ability.

Breeding from trapnested stock. — Where fowls are trapnested, it is possible to engage in a much more elaborate series of breeding operations, with the idea in view of improving the egg production. It must be stated, however, that trapnesting is a rather laborious process and will usually be found to be too expensive for the average

Fig. 25. White Plymouth Rock hen 740. This hen was not trapped until January 26 of her pullet year. From this time on she laid 176 eggs, 84 eggs being laid in 92 consecutive days and 109 in 122 consecutive days. (Photograph from the Bureau of Animal Industry, United States Department of Agriculture.)

person to attempt to follow. At the same time it must be recognized that the interest in breeding for improved egg production is keen at the present time, and that the demand for stock from high producing lines is good and will probably be better in the future. It is likely, therefore, that it will pay many producers to go to the extra labor and expense of trapping their birds in order to be able to supply this demand for breeding stock out of known high producing individuals. Since there is no practical means which have yet been devised of certifying to or absolutely authenticating the records obtained by breeders in their own yards, it is a foregone conclusion that the integrity and reputation of a breeder must be of the best if he is to succeed in this business of producing high record breeding stock for sale. It is also possible, as will be explained later, for the breeder to trapnest to some extent and to be able to supply for his own needs, at any rate, birds from known high producers for his breeding operations.

Fig. 26. Trap nest in use. (Photograph from the Bureau of Animal Industry, United States Department of Agriculture.)

Where the fowls are trapnested, it is necessary, if the breeding is to be carried on with individuals whose ancestry is definitely known on both sides, that the chicks should be pedigreed. For directions as to pedigreeing see page 40. The pedigree hatching and rearing of the chicks makes it possible, therefore, to have on hand a number of fowls which can be used in the future breeding operations with absolute certainty as to the performance of their ancestors with respect to egg production, both on the male and female sides.

The detail of the matings which should be made is a matter which must depend very largely upon the judgment of the breeder himself. He must, however, keep in mind that all of the birds bred must be strong and healthy and must possess in so far as he can determine vigor to an unusual degree. The breeder must never allow mere high egg production in the ancestry to overbalance his good judgment and allow him to succumb to the temptation of using a bird of relatively low vitality and vigor.

The usual method of mating has for its purpose the combining in the pedigree of the offspring blood of high producing strains, both on the male and female sides. Again, however, it is necessary to call attention to the fact that too close and too long continued inbreeding must be avoided. While it does not necessarily follow that the sons of high producing hens will prove to be breeders whose daughters will all be good producers, it is certainly true that the daughters of males whose mothers were high producers have a much better chance of becoming in turn high producers than do those out of a male whose mother was a poor producer. As the pullets from the different matings mature and are put in the pens and trapnested, it becomes possible to check up their performance and to determine what individuals, either male or female, have been prepotent breeders with respect to the egg production of their daughters. Needless to say, if any indiviudals are discovered which show marked prepotency in this respect, they should be retained in the flock and bred just so long as they are useful. It is also, of course, common sense to use the sons of such prepotent individuals in other matings so as to test out their prepotency in turn. If prepotent individuals are discovered of each sex, it is by all means desirable to mate these together, keeping in mind all the time that too close and long continued inbreeding had best be avoided.

Dr. Raymond Pearl, as a result of his experimental work at the Maine experiment station, outlines the following plan for the practical breeder who is trapnesting his birds to follow in breeding for increased egg production :*

1. "Selection of all breeding birds first on the basis of constitutional vigor and vitality."

2. "The use as breeders of such females only as have shown themselves by trapnest records to be high producers."

3. "The use as breeders of such males only as are known to be the sons of high producing dams."

4. "The use of a pedigree system whereby it will be possible at least to tell what individual male bird was the sire of any particular female. This amounts to a pen pedigree system which can be operated by the use of a toe punch."

5. "The making at first of as many different matings as possible. This means the use of as many different male birds as possible, which will further imply small matings with only comparatively few females to a single male."

6. "Continued, though not too narrow, inbreeding or line breeding of those lines in which the trapnest records show a preponderant number of daughters to be high producers. One should not discard all but the single best line, but should keep a half dozen at least of the lines which throw the highest proportions of high layers, breeding each line within itself."

Because trapnesting, if done for a large flock, is not only a laborious, but an expensive process, relatively few breeders find it practical for them to attempt to trapnest all their birds. It is, however, possible to combine the system of breeding so that they will have the advantages of some trapnested stock, and will be able to check the results in the offspring by keeping pen records. For example, it is possible to trapnest two or three pens of pullets, and when high producing individuals are discovered among these birds they can be used for breeding, the chicks from them can be pedigreed, the young males in turn being used on other breeding pens, and the pullets placed in the trapnest pens for

*Maine Agricultural Experiment Station Bulletin No. 231.

TRAP NEST RECORD OF HEN No. 691.

1914	1	2	3	4	5	6	7	8	9	10	11	12	13	14	15	16	17	18	19	20	21	22	23	24	25	26	27	28	29	30	31		
Nov.																																0	0
Dec.											/		/	/		/	/	/		/		/		/	/	/		/	/	/		14	14
1915 Jan.	/	/		/	/		/	/	/		/	/		/	/	/		/	/			/	/	/	/		/	/	/		/	22	36
Feb.	/		/	/		/		/		/	/	/	/		/	/	/	/		/	/	/		/	/	/	/	/				21	57
Mar.	/		/	/	/	/	/	/	/	/		/	/	/	/		/	/	/		/	/	/		/	/	/	/		/	/	25	82
Apr.	/	/	/	/	/	/		/	/	/	/	/		/	/	/	/		/	/	/	/	/	/	/	/	/	/		/		26	108
May	/	/	/	/	/	/		/	/	/		/	/	/	/	/	/	/	N	N	/		N		B			O				17	125
June							/	/	/	/		/	/	/	/	/	/	/	/	/	/	/	/		/	/	/	N	/			20	145
July		/		N		N				B			O									N	N	/	/	/	/		/	/	/	8	153
Aug.		/	/	/		/	/	/		/	/	/		/	/	/		/	/	/		S	/	/		/	/	/		/		21	174
Sept.	/			/	/		/		/	/		/	/		N		/	/	/		/	/		/		/	/		/	/		18	192
Oct.		/	/		/		/		/		/		/		/		/		/		/											11	203

N = Indicates that hen went on nest and did not lay.
B = Indicates that hen went Broody.
O = Indicates that hen was released from broody coop.
S = Indicates Small egg.

Fig. 27. Trap nest record for an individual hen. The monthly records are copied from the pen trap nest records Fig. 18 and combined on this card. (From the Bureau of Animal Industry, United States Department of Agriculture.)

testing. Then, by housing or penning together the pullets from these different males, which are used on non-trap-nested stock, a pen record of egg production can be kept and these males tested with respect to their prepotency. In this way breeding stock is secured from high producing lines without going to the expense of trapping the whole flock or keeping individual pedigree records of all the chicks hatched and cockerels from high producing lines are available for use on the general or non-trapnested pens.

Period of trapping.—Where the sale of stock from record hens is contemplated, it will be necessary to trap for the entire year, as the demand is for records on a yearly basis. Usually the pullet year record is the highest year's record and unless it is desired to work for a reputation for long distance layers, that is, hens which keep up their good production for several years, it is unnecessary to trap beyond the end of the pullet laying year. However, it is usually good policy to continue to trap all hens which produce 200 eggs or better in their pullet year, for a long distance record is a good selling point. In order to cut down the number of pullets which are being trapped, it is usually possible by March 1 or before to eliminate a large proportion which obviously have no chance to make a good record. Dr. Pearl in his work considered any pullet which laid 30 eggs or more from November 1 to March 1 to be a good winter layer and a profitable and likely bird to keep. In general, it may be said that a pullet which lays 50 to 60 eggs from November 1 to March 1 has a good chance to make a record of 200 eggs or better.

CHAPTER IV

THE AMERICAN CLASS

The Plymouth Rock

The Plymouth Rock is a good sized, upstanding fowl, with body of good length, good breadth and depth. There is a considerable tendency for individuals of both sexes to be too small, and this can only be corrected by selecting fowls of standard weight as breeders. Birds should not, on the other hand, be used which are more than one or two pounds over weight, as such unusually large birds almost invariably get away from good Plymouth Rock type.

The Plymouth Rock should not be stilty, that is, too long in shank. The bird wants to stand strongly on legs set well apart. There should be no tendency to a knock-kneed condition, but this is by no means rare and is quite troublesome. See Fig. 28. The breast should be full and well rounded, not flat. Males are particularly likely to be lacking in breast. See Fig. 29.

Fig. 28. A knock-kneed male. (Photograph from the Bureau of Animal Industry, United States Department of Agriculture.)

A long keel is desired and the body should have good width. Birds with narrow bodies should be avoided as

69

breeders, since they are not typical Plymouth Rocks. A body set too low on legs, too deep from back to hock, or, in other words, body approaching too closely the Orpington in type, or with a tendency to be baggy, should be avoided.

Fig. 29. A stilty Barred Plymouth Rock male with flat breast. (Photograph from the Bureau of Animal Industry, United States Department of Agriculture.)

The comb of the male should be fairly small, with a well-turned blade and five points. The comb of the male follows the curve of the head in general outline, but the blade should not follow the neck too closely. While five points are desired, four points, because of the relatively small size of the comb, are not bad, as they tend to form a symmetrical comb. In some cases, a six-point comb may prove to be symmetrical also. It is better to use a male with a four-point comb than one with six points, especially when the females to which he is to be bred carry combs with an extra number of points. The comb should be thick at the base and be strongly erect. It should not extend too far forward over the upper mandible or bill.

The comb of the female should be small and erect, like the male comb, but smaller. A comb in which the top of the points form a nearly straight line, or is a little rounded, is considered good in shape. The combs of Plymouth Rocks run quite good and are not generally considered very troublesome. The following are some of the more common faults: too large, too many points, side sprigs, double points and thumb marks, wrinkles or corrugations.

In the shape of head, the important thing is to see that the head is broad, thick and of good length, but not extremely long and snaky or crow-headed. See Fig. 8. The broad, deep, full skull is generally considered to denote stamina, and crow-headedness a lack of it.

The eye in each variety should be full of life and fire. Light or green eyes must be avoided.

The ear lobe should, of course, be red in both sexes. The trouble which may occur here is the appearance of white. Distinction must be made between positive white and the paleness, which may be due to lack of condition. If possible to avoid it, do not breed from an individual showing positive white, but the white which borders on paleness is not so serious. It must be remembered also that white or paleness sometimes develops with age. This, of course, is not as serious as where it occurs in young birds, and in consequence knowledge as to the color of the ear lobe of a bird while it was still a cockerel or pullet is valuable when mating.

The neck of both sexes should be well arched. In the male it is long and full, with abundant hackle, while in the female it is of medium length.

The back is one of the most important sections of this breed. In order to be a good typical Plymouth Rock, the back must be nearly flat at the shoulders, long and broad throughout. As nearly as possible, the same breadth which is present at the shoulders should be carried the entire length of back clear to the base of the tail. There should be no tendency to narrow up at the tail.

The tail is also important, especially so because of its influence upon the shape of the back. The tail in both sexes is carried moderately low. There should be no angle between the back and tail, but these sections should be joined by a good sweeping curve, the back and tail blending in together. It is of vital importance that the tail be well spread. By this is meant that the tail shall have good breadth when v i e w e d from above, and shall be arched. If the tail lacks this spread and b r e a d t h, the effect will be that of the back narrowing up at the base of the tail. If the tail has a g o o d s p r e a d, however, then it can carry out without b r e a k the b r e a d t h and l i n e s of the b a c k. Sometimes the tail of the male, especially if v e r y w i d e, will be split or divided, but this is undesirable. Split tails in males are more or less common and undesirable in all breeds having widespread tails.

Fig. 30. White Plymouth Rock male showing split tail. (Photograph from the Bureau of Animal Industry, United States Department of Agriculture.)

The wings should be carried properly folded and held strongly up in place. In a properly folded and held wing, the point of the wing points to, and is in line with, the vent. Frequently the point of wing is held up too high toward the saddle. Quite often the wing in this breed is not held up properly, but hangs down partly unfolded. Such a wing is called a slipped or split wing. See Fig. 8. Slipped wings are especially prevalent in the white variety. Breeders should be selected which are free from this defect if possible, for it seems to breed strongly and is difficult to get out of a flock, once it has been introduced.

Sometimes the flight feathers do not grow normally, but show a twisted condition. See Fig. 31. Twisted wings are especially prevalent in Barred Plymouth Rocks. This is quite a serious defect. Constitutional weakness is apt to be associated with twisted wings. Birds with twisted wings usually have long wings

Fig. 31. Barred Plymouth Rock male showing twisted wing. (Photograph from the Bureau of Animal Industry, United States Department of Agriculture.)

poorly fleshed. A well-fleshed wing seldom shows twisted feathers.

A good yellow leg color is desired in all varieties, but is more or less difficult to obtain in some varieties, particularly in the females. Green peppering is likely to show in the White Plymouth Rock, a dark cast in the females of the

Partridge variety, dark spots or a green shade in the females of the Barred variety, and a green shade in the exhibition Barred Plymouth Rock male. Often Barred Plymouth Rock pullets which show dark shading on the shanks molt in with a fine yellow color of shanks as hens. While it is more or less difficult to secure clear yellow legs in the cases mentioned above, it is possible to do so. This particular will undoubtedly be improved vastly, so that it will pay breeders to do what they can toward this end.

Barred Plymouth Rock females often show a dark shading in beak. While this is not considered a serious defect, a clear yellow is preferred.

The Plymouth Rock should have large, heavy, well-muscled thighs and shanks of good length. Too long shanks must, however, be avoided, as they are apt to go with flat breast. Too long shanks, knock-knees and long or crooked toes are often found in birds of weak constitution. The toes are sometimes crooked on one or both feet. This, of course, detracts from the appearance of the bird and consequently injures its show quality considerably. Stubs, small feathers or down sometimes occur, but the breed runs pretty good in this respect. In mating Plymouth Rocks the following are defects more or less common to the breed which must be guarded against so far as possible:

Too large comb; too many points to comb; side sprigs; double points to comb; thumb marks; too small size; too stilty or too long-legged; too deep, low set bodies approaching the Orpington in type; flat or deficient breast; knock-knees; narrow bodies; crooked toes; tail too high; split tail; tail not well spread; slipped or split wings; twisted wings; dark spots, dark or green shade or green peppering in shanks, particularly of females; stubs; light or green eye; white in ear lobe; crow or snaky head; back tending to narrow in at base of tail, thus not carrying the width of back at shoulders throughout the length of back; angle where back and tail join.

The Barred Plymouth Rock

This variety has been bred for a long time, almost exclusively under the double mating system. As a result, the cockerel and pullet lines of blood are very definitely fixed, and it is of the greatest importance for success that the birds used in the matings should be of the right breeding. It is inadvisable under most circumstances to cross the cockerel and pullet lines. So important is the breeding back of the birds, and this, of course, holds true of all line-bred strains or families of any variety or breed, that a bird of the right breeding should often be selected for a mating in preference to one of far better quality but not of the right breeding.

It is also well to note here that solid black feathers, feathers partly black and partly barred, and feathers similar to the Dark Brahma hackle feathers, will be frequently found throughout the plumage of both male and female Barred Plymouth Rocks. The occurrence of these feathers does not indicate any impurity of blood, nor are they considered serious defects unless found in very large numbers. These off-colored feathers, and in particular the solid blacks, are numerous in the plumage of many of the very finest specimens, and are, of course, pulled out before the birds are shown. See Figs. 10 and 11.

The breeding of Barred Plymouth Rocks is such a specialized business that it seems best to set forth the methods of mating as given by some of the individual breeders.

Cockerel mating.—A breeder of cockerel-bred birds describes his method of mating as follows: The male should be the finest quality of exhibition male possible, provided he is descended from a long line of high class exhibition males. He should be of good size, that is, of standard weight, but should not be more than a pound or two above standard. He should be a fairly quick growing bird, and should be fully matured in form and feather at an age of not less than seven months, as there is a tendency for some strains of this variety to be slow in maturing.

The surface color of the male should be clean and even, with an underbarring consistent with the brilliant surface color. It is most important that the surface color be even all over—that is, that the hackle, back, wing bows, saddle and tail coverts should be one even shade of color. This evenness of color can be easily tested by pressing the bird's hackle back on the wing bows and saddle, when any variation of color is at once perceived. In the cockerel-bred males more or less metallic will often be found on the wing bars, tail coverts and sickle feathers. This metallic shows as a greenish cast in the dark bars of the feathers and is a defect which should be guarded against in selecting males both for breeding and exhibition purposes. Under no circumstances use a male with a smutty colored surface—that is, when the dark and white bars are not clearly defined and distinct, but the dark tends to run into the white bar. It is a mistake to try to clear up this color in the offspring by using a bright colored female. The white bar should be clean, distinct and snappy, and the black bar free from a brown edging. The contrast between the black and the white should be as distinct as possible, and the line of demarcation between the two bars very definite and clean-cut. The under color does not necessarily have to be heavily barred, but the barring should be well defined clear to the skin. See Figs. 32 and 33. The wing-flights and secondaries should be straight and distinctly marbled, and the barring should run straight across the wing. If the barring in the wing of a male shows pronounced black and white, this denotes probable strength of barring throughout the entire plumage. Good strength of barring should also be shown in the fluff, as this is generally one of the weak sections. A bird strong in wing and fluff barring, and even in surface color, will be strong in the underbarring of other sections. In general, in selecting a male to head the pen, pick one which is, as nearly as pos-

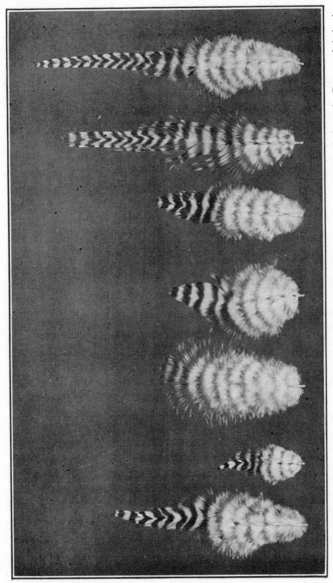

Fig. 32. Well barred feathers from a cockerel bred Barred Plymouth Rock male. (Photograph from the Bureau of Animal Industry, United States Department of Agriculture.)

77

sible, in color, type and barring, exactly what you wish to see in the progeny.

In selecting females for this mating, first be sure that the breeding is right—that is, that they are descended from a long line of high class exhibition males. The female should have a tail carriage slightly below standard requirements, and the tail should be well spread. She should also be broad and full feathered over the back, and this should be carried well up on to the tail, even to the extent of showing a slight cushion. It is from such females that the good, broad-backed males with full, long feathered saddles are secured. The comb should be low, evenly serrated, and have four, five or six points, which should be well defined in contour.

In color the female should be clean black and white on the surface, with no sign of smut. It is, however, very difficult to obtain absolute freedom from smut. The black bar on the surface should be two or three, or even four times as wide as the white bar, and the surface color should be even in every section. The underbarring should be narrow and well defined clear to the skin in all sections, with no sign of smutty color. See Fig. 33. The barring in the hackle should be especially straight across the feather in order to produce the finest hackles in the exhibition males. The black and white marbling of the wing should be very clear. It is very important that the tail barring be straight across the feathers. Some of the best individual females for the cockerel mating will have some entirely black feathers even in the wing flights. This denotes an abundance of pigment, and will produce stronger underbarring in the progeny. The barring in the fluff should be good, in fact, just as good as in the back. Cockerel-bred females are usually darker colored on the legs and beaks than exhibition females, but this does not hurt their breeding value for producing exhibition males.

In both sexes the barring on the breast should be straight and with a sharp contrast in color. If the barring is weak

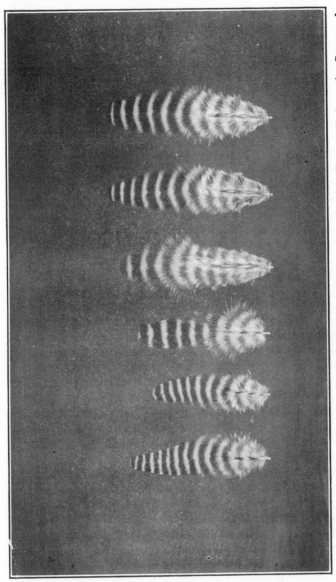

Fig. 33. Well barred feathers from a cockerel bred Barred Plymouth Rock female. (Photograph from the Bureau of Animal Industry, United States Department of Agriculture.)

79

or poor in this section the barring on the lower breast and toward the thighs is apt to run pretty wide.

In this mating the following defects must be guarded against, so far as possible: Too large combs; green eyes; under color not well defined; white or cotton under color; green legs in exhibition males; double points to comb; smutty feathers in wing bows; slipped wings; metallic on the wing bars, tail coverts and sickle feathers of males.

Pullet mating.—A breeder of the pullet line gives the following as his method of mating: The male to head this mating should be rather long in body, should carry the tail a trifle lower than the standard male, and have the width of back carried out proportionately in the width of tail. The tail should be fairly well spread, but the sickles and coverts do not necessarily need to be as fully furnished as in a show male. The breast should be wide and full, a flat breast being guarded against. The legs should be strong and set well apart, with no tendency toward knock-knees. The comb may be a trifle smaller than that of the standard or exhibition male. A very strong, bright red eye is desired. The ideal color of the male for this mating is the same as that of the ideal exhibition female, but the barring of the male runs much narrower than in the exhibition female, and this greater width of the dark bar in the female causes her to appear to be of darker shade than the pullet-bred male. While the male should be straight pullet-bred, even in such a line males will be found ranging widely in color from a medium strong color to a very light color. This difference in color should be due entirely to the difference in the width of the dark bar, as the dark bar gives the color, and should not be due to any difference in the color of the dark bar itself. Have the white bar as white as possible in both sexes, as the clearness and distinctness of the white bar adds materially to the breeding value of the bird. Provided the dark bar is ideal in color, but still the birds are of several shades of color due to the difference in the width

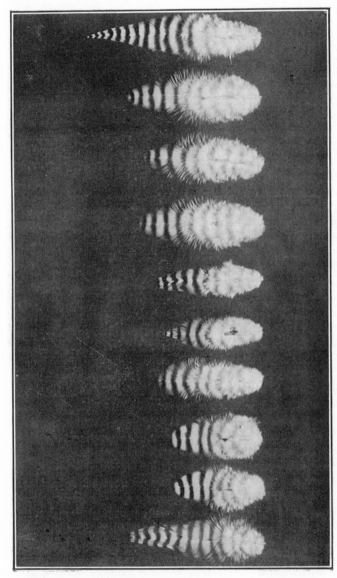

Fig. 34. Well barred feathers from a pullet bred Barred Plymouth Rock male. (Photograph from the Bureau of Animal Industry, United States Department of Agriculture.)

81

of the dark bar, the success in breeding will depend upon the ability to combine the different shades of the male and female. The male showing the narrowest dark bars should be mated with the females showing the wider dark bars, and vice versa. This matter of the shade of the birds or the width of the dark bar practically resolves itself into a mating of feathers. Select a male that shows as clear and bright contrast between the light and dark bars as possible. The two colors should be clear and distinct, the white as white and the black as black as possible. See Fig. 34. This distinctness can often be found in the highest degree in the wing bows or shoulders of birds of both sexes, and can be used as a guide to color. Avoid any brown or grayish cast shading off the dark bar into the white. The line between the two bars should be clear and distinct.

The hackle of the male often shows a tendency to be a little wider or more open in barring than other sections, and a male good in other respects and showing this openness in hackle should not be discarded, as he may prove to be a most valuable breeder. If there is too fine a barring in the hackle of the male, this is apt to show in the female offspring as a dark or smutty appearing bar.

If the dark bar is more than equal to the width of the white bar in the under color of the back, which, however, is not often the case, the female offspring generally come indistinct and with slaty under color. Often cockerels, when getting their adult plumage, show a yellow cast in hackle, back, wing bows and saddle, but principally in hackle, which disappears when the bird is fully finished. This yellow cast is often mistaken for brassiness in the Barred Plymouth Rock male. The wing flights should show a good, clear, distinct, straight barring. Birds weak in this respect should not be used in the breeding pen if possible to get along without them, and if used, great care should be taken to offset the defect in the selection of females for the mating. Birds showing a good, strong fluff barring, both in surface and

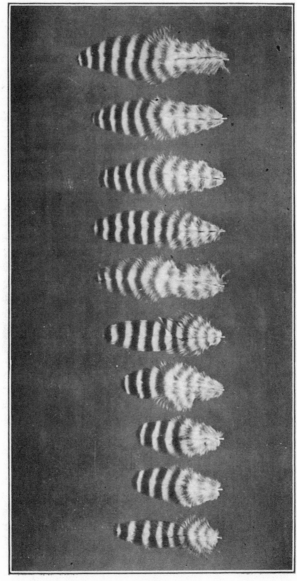

Fig. 35. Well barred feathers from a pullet bred Barred Plymouth Rock female. (Photograph from the Bureau of Animal Industry, United States Department of Agriculture.)

under color, are as a rule considered very desirable as breeders.

The females for the pullet mating should be as near standard as possible, but if bred along the right lines may be used when they vary all the way from the narrow dark bar to the wide dark bar—that is, are too light or too dark—the weakness of the barring of the females being offset by the selection of a male strong in those points. The barring of the females must, however, be clear and distinct—that is, the black as intense as possible and the white clear. See Fig. 35. Avoid a brownish cast or dull color in the black bar. While the standard calls for back feathers ending in a black tip, they are frequently lacking in this respect, especially in young birds, and this is a serious defect. Pullets often show good black-tipped feathers on wing bows and breast, and cockerels on hackle and back. Pullets generally show the most white tips in the back. However, young birds generally show improvement in this respect with age.

As in the case of the male, the barring of the wing flights should be straight, clear and distinct, and females weak in this respect should not be used if it is possible to avoid doing so. Even the finest of exhibition females rarely show solid yellow beaks and legs that are absolutely free from any foreign color. This is especially true of the beaks. Dark streaks or shading are present in the beaks and dark spots on the legs. Often pullets showing some dark in shank color molt in a fine yellow as hens. The clear yellow color is most desirable, however, and should be selected for where possible without sacrificing good color and barring.

The barring in both the male and female often is better in adult birds than in cockerels and pullets. Therefore do not discard young birds showing promise until they have fully matured. Early hatched birds of both matings sometimes seem to show a brown, which is due to the fading incident to long exposure to the sun. The later hatched birds do not show this as much, since their mature plumage

Fig. 36. Cockerel bred Barred Plymouth Rock cockerel on the left and pullet bred cockerel on the right. Notice the difference in the relative widths of the black and white bars in the two birds and the difference in color of body plumage generally which this causes. (Photograph from the Bureau of Animal Industry, United States Department of Agriculture.)

Fig. 37. Cockerel bred Barred Plymouth Rock pullet on the left and pullet bred pullet on the right. The difference in the relative widths of the black and white bars in the two birds is responsible for the difference in the shade of the plumage as a whole. (Photograph from the Bureau of Animal Industry, United States Department of Agriculture.)

is not subjected to as much sun. This is responsible for the belief rather commonly held that later hatched birds are superior in color to early hatched birds.

The following defects must be guarded against, so far as possible, in this mating:

Green or light eyes; weak, flat breasts, especially in males; knock-knees and crooked toes, especially in males; green shade or pale yellow legs in both sexes; dark spots on shanks in females; slipped and twisted wings; crooked backs (this defect is not necessarily transmitted to the offspring); solid black feathers in wings (while it is preferable to use breeders, if possible, which are free from this defect, many high class specimens showing it may prove to be very valuable breeders, as the defect is not necessarily transmitted to the offspring. Solid black feathers in the wings may not be present in young birds, but may come in with age); smoky or mossy feathers in back of females; uneven tipped feathers in back of females, that is, feathers some of which are tipped with white and some with black, showing no uniformity in the tipping.

In addition to the defects to be guarded against, as given under the cockerel and pullet matings, the defects listed for the breed in general (page 74) should also be consulted.

The White Plymouth Rock

This variety is probably the best in type and the most uniform in this respect of any of the Rocks. In breeding this variety it is usual to employ a single or standard mating, or to use a standard male with two types of females.

The male should be as near standard as possible. The color of the male, and of the females as well, should be a good, clear white, with no trace of creaminess in the quills. Often a small amount of black or dark ticking will appear in the wing and tail feathers. This should be guarded against, as it is likely to prove troublesome. Also avoid gray in the flight feathers. Special importance must be attached to the back of the male. The breadth of back must be carried throughout and there must be no tendency toward narrowing in at the tail. The tail should be well furnished, well spread, and carried rather low. Carriage of tail lower than called for by the standard seems to find favor in the show room, but breeders should be selected which are standard in tail carriage unless the females used show a tendency toward too high tails, when a male with lower carried tail should be used. In some males, the tail splits in the middle and the sickles fall between. This is undesirable and should be guarded against. See Fig. 30.

To mate with this male for the production of exhibition cockerels especially, select females which show a little cushion and which are a trifle shorter in back and a trifle higher on legs than exhibition females. They should also be full in fluff, the cushion and fluff forming a circle when viewed from the rear. The cushion and full fluff tend to give a nicer finished back and tail on the cockerels.

For the production of exhibition females, and of males as well, select females as near to standard as possible, showing a full, deep breast and a short, well-spread tail, carried nearly on a straight line with the back, and with as little narrowing of the back at the tail as possible. By well-spread tail is meant a spread from side to side, as viewed from above. The tail should be slightly arched, but not duck shaped—that is, not flat on top.

Breeders of both sexes should be up to or slightly above the standard weight, as it is difficult to get back standard size once it is lost.

In addition to the general defects for the breed as a whole (page 74), the following defects must be especially guarded against, so far as possible, in this variety:

Black or dark ticking; creaminess in quills; horn color at butt of quills of flight feathers; slipped wings; green peppering in the shanks; stubs; split tails.

The Buff Plymouth Rock

In this variety there is a tendency for the males especially to be set too high on legs, that is, to be stilty, and also to be lacking or flat in breast.   Combs which are too large are also very frequent.

In mating this variety it is not necessary to resort to double mating, as high class standard specimens of both sexes can be produced from a single or standard mating.

In selecting breeders for a standard mating, choose birds of both sexes which approach the standard as nearly as possible, both in color and type.   Evenness of color in both male and female is especially desirable, and each bird should be one even shade from head to tail.   The color of the male will, of course, be somewhat heavier than that of the females, but evenness must be emphasized in each bird.   The general surface color of the females selected should be the same shade of buff as the breast of the male.   Select in so far as possible birds which are free from white, black or a peppering of either color in the wing flights and the main tail feathers of both sexes, and in the sickles of males.   Of these two foreign colors, black and white, black is generally considered the less harmful.   There is a decided tendency for the females to be mealy on the shoulders.   By mealiness is meant an unevenness of color which causes the surface of the feathers to appear as though sprinkled with some lighter colored substance, such as meal.   In fact, mealiness is one of the greatest faults in all buff females.

White in the under color, especially of the hackle, back and saddle, and at the root of the tail of males, is a common and troublesome defect, and one which is apt to develop or to increase with age.   Therefore cockerels sound in this respect are more valuable as breeders than those which show it even slightly.   Old males sound in this particular are especially valuable breeders of under color.   There is also a tendency for the quills of the feathers to be very light buff or white close to the body in both sexes. This is especially true of the quills of the main wing and tail feathers, which seldom if ever show a good buff clear to the skin. In some specimens there is a tendency toward a darker colored edging or lacing on some of the feathers, which should be avoided. Some birds show a pinkish cast to the buff under color, and such specimens are most likely to be sound in surface color and to produce offspring sound in surface and under color.

While double mating is not necessary in order to produce standard specimens of both sexes, it is often resorted to. When this is done the matings should be selected as follows:

Cockerel mating.—The male should be standard and absolutely sound in color, that is, free from any black or white in wings or tail and from a reddish cast on hackle, back, wing bows and saddle, or from a red edging to the front of the hackle.

The females must be bred from a cockerel line and sound in color. The shade of their general surface color should be slightly darker than that of the male's breast. This is a somewhat darker shade than that ordinarily sought for in exhibition females. A very important point is for the females to have as rich and deep an under color as possible, extending clear to the skin. In many instances such females will have almost as good an under color as the male. This is the mating depended upon to produce exhibition males, while the females are usually too dark in surface for exhibition. Occasionally, however, a female is produced from such a mating with light enough surface color to be suitable for exhibition, and such females are usually very good because strong in under color.

Pullet mating.—The male used must be from a pullet line and may not be as even colored as in the cockerel mating. In fact, a deeper color on the wing bows and shoulders tends to offset any inclination to mealiness in the surface of the females, which is quite a troublesome defect.

The breast of the male should be of the same shade as the general surface color of the exhibition female, or, in other words, of the females used in this mating. The male's breast should be absolutely free from shafting, for, if any is present, the pullets from this mating are almost sure to show some shafting throughout, which is a serious defect. In surface color the male should be as nearly sound as possible, but may be lighter both in surface and under color than the standard, if out of a high class exhibition female. The females of this mating should be as near standard as possible and free from shafting, mealiness or patchiness. By patchiness is meant the presence of irregular spots or patches of different shades of color. The under color of these females need not be as deep as that of the cockerel-bred females, but should be buff with a density in proportion to the surface color.

The following are defects which must be especially avoided, in so far as possible, in this variety, in addition to those defects which are common to all the varieties of the breed (see page 74): Too large, coarse combs; thumb marks; lopped combs; white, black or a peppering of either in wing flights or in main tail feathers of both sexes; white edging to the sickle feathers of males; shafting, mealiness and patchiness in females; unevenness of color of the hackle, back, wing bows and saddle of males; willow or light-colored shanks, especially in females; white in under color; white or very light buff quills close to the body; feathers tending to be laced.

The Silver Penciled Plymouth Rock

Both the single and double mating systems are used in breeding this variety. In double mating, select the breeders as follows:

Cockerel mating.—The male should be standard in color, with sharp, clear laced saddle and hackle. The breast and body color should be black. The females of this mating should be rather dark in color, with black penciling and clear, sharply striped neck hackles.

Pullet mating.—The male should show some frosting and white splashing on breast, thighs and fluff, but not up under throat. If a male is used with too much light color or mottling of white up under the throat, the females are apt to run too light in that section. Occasional males even show some lacing of gray on the sides of the body, and these are especially valuable pullet breeders. The females should be standard in color.

A single or standard mating is sometimes used with considerable success if the breeder prefers the heavier black penciling in the females. For such a mating these black penciled females are mated to a standard male.

Another single mating which may be used is to select birds of both sexes which are as near the standard as possible. It is important if exhibition birds of both sexes are to be secured in the offspring from this mating that the male used should be out of a well-penciled female.

In addition to the defects for the breed as a whole, the following common breeding defects of this variety must be avoided in so far as possible:

Red on wings of males; solid white in flight and main wing feathers of males, other than the white edging of the primaries and secondaries called for by the standard; brassy backs of males; inclination to brownish cast instead of gray in females.

For additional information on mating for Silver Penciled color, the reader should refer to the Silver Penciled Wyandotte (page 113), and to the Dark Brahma (page 140).

The Partridge Plymouth Rock

In breeding this variety some breeders use the single or standard mating, while others resort to a double mating.

In the single mating, select a male with rich, medium red color, neither the light orange that frequently occurs nor a red so dark that it is difficult to distinguish the

Fig. 38. Well marked Partridge Plymouth Rock feathers. M indicates male and F female. (Photograph from the Bureau of Animal Industry, United States Department of Agriculture.)

stripe of the hackle and saddle. It is important that the red top color of the male be even in shade. It is also important that the striping in hackle and saddle be a lustrous greenish black and that the black does not extend through the end of the feathers, but that the edging of red extends clear around the end of the feather. The hackle and saddle should be free from white or cotton under color, and this is especially important in the neck. The under color of the body should be slate of medium shade. The breast and the fluff should be solid black. Some breeders claim that a male with brown in the fluff will breed better females, but this is denied by others. The male should have clear yellow legs.

The females for the single or standard mating should be mahogany in color and of even shade. The penciling should be regular in all sections of the body, from well up on the throat to the tail and fluff. The shanks of the females are very apt to show dark, but those as near a clear yellow as possible should be selected.

When double mating is employed, the matings are made as follows:

Pullet mating.—The male should be lighter in the red sections than the exhibition male, approaching a lemon shade. He should also show a red ticking on the breast, and a slight amount of red in the fluff. He should show a lighter under color than the exhibition male in all sections except the hackle and tail, where it should be as dark as possible.

The females for this mating should be as near standard as possible.

Cockerel mating.—The male should show a fine, clear, cherry red of even shade in all red sections. The striping in hackle, back and saddle should be clear and sharp, the black showing a good sheen, and not extending through the end of the feathers.

The females for this mating should be of standard color, but showing a distinct greenish sheen.

In mating this variety, the following defects in addition to those common to the breed (page 74) should be guarded against in so far as possible:

Light eyes; black striping running through end of hackle and saddle feathers of males so that there is no red edging showing clear around the end of the feather; brown shafting in the hackle and saddle striping of males; white in tail and wings, especially in males; dark legs, particularly in females; stubs; too light colored females with lemon hackles; stippling in the tail coverts of females.

For additional information with respect to mating for Partridge color, the reader is referred to the Partridge Cochin (page 147).

The Columbian Plymouth Rock

This variety has a tendency to fail in type, often being too rangy and carrying the tail too high. There is also a tendency for the fowls to run too light in color—that is, the black sections are apt to be faded instead of the good intense black which is desired in order to show a strong contrast to the white. This is particularly true of the hackles of the males. To offset this tendency, it is customary to use breeders which have a dark slate under color.

In mating this variety, both the single or standard mating and the double mating are employed. The single mating is more common and is advised.

In the single mating, select birds of both sexes which have hackle feathers as near as possible clear black, with white edges running all the way around the point. The color of these feathers should be a positive black and white. The tail should be a solid black. In the females the tail coverts should have a white lacing, while in the males, the lesser tail coverts and some of the saddle feathers should be black with white lacing. The surface color of both sexes should be a clean, distinct black and white, with a bluish white under color. The under color may, in fact, be quite dark, even to the extent of the fluff showing a bluish tinge on the surface. Formerly, in order to increase the black of the wing feathers, there was a tendency to use females showing black on the surface of the back. Such females or males showing too heavy a striping in hackle, or birds of either sex showing black in the surface of the feathers on the sides of the fluff or the body feathers just in front of fluff should not, however, be used. In general in this mating, keep away from birds as breeders showing a pure white under color, and use those showing a pure white surface color.

Where the double mating system is used, the following matings are made:

Cockerel mating.—Select a standard colored male with a good, strong green sheen in all black parts. The hackle should be clearly and distinctly striped clear to the skin, while the saddle is not heavily laced, the striping running about one-half way down in the under color. The back should be absolutely clean in color. The main tail feathers should be a solid black, the black running clear to the base of the feathers. The fluff should be white, ending with white at the skin.

The females should be of standard type. The feathers should be broad. The hackle lacing should come well around in front of the neck, while the back should show no ticking in surface. The under color of back should be a strong, bluish white. The wings should show strong color, the black and white being distinct and clearly defined. Good, narrow, sharp lacing is desired in the tail coverts.

Pullet mating.—The male should be of good type and

head points. He should be more heavily colored in all sections, including under color, than called for by the standard. The saddle and hackle should both be long and flowing and heavily striped. The tail should be profusely furnished with coverts which have a distinct and narrow lacing running the full length of the feather and free from brown color. It is very important to have strong black coloring in the wing primaries and secondaries, the black and white being distinct and free from peppering. The females for this mating should be lighter than standard, a white under color of fluff being desired. The lacing of the hackle and tail coverts should, however, be as distinct and well defined as possible.

The following defects, in addition to those common to the breed (page 74), should be especially guarded against in so far as possible in breeding this variety:

Too large comb; too many points to comb; side sprigs; too rangy in type; tails carried too high; black in the surface of back of female; birds too small in size; stubs and down; black in surface of feathers at sides of fluff and the body feathers just in front of fluff.

Additional information as to breeding Columbian color will be found under the Light Brahma (page 134), the color scheme of which is identical with that of the Columbian varieties. It must be kept in mind, however, that the Columbian Plymouth Rock is a newer variety than the Light Brahma and that it has not as yet been developed to the same degree of perfection.

The Wyandotte

The Wyandotte is a well-rounded bird and therefore a bird of curves. There should be no angles in the outline and no straight lines. The Wyandotte must be a well-balanced fowl, and to achieve this balance the legs must be under the center of the bird and the distance from the top of the back to the bottom of the feet should equal that from the breast to the end of the tail. The body should be horizontal and a good depth of body must be maintained, as there is a tendency for some birds to lack in this particular. The Wyandotte size should be maintained, but extra large and coarse birds are not desirable as breeders, since they are almost sure to lose the roundness a n d compactness of s h a p e desired. N e i t h e r are b i r d s which tend to be too r a n g y or too narrow - bodied desirable. S e e Fig. 39. Breadth of b o d y and b a c k is neces- sary in order to maintain t h e meat - carrying carcass.

Fig. 39. Buff Wyandotte male, which is too rangy, too flat in back, and which has a comb too large and coarse. (Photograph from the Bureau of Animal Industry, United States Department of Agriculture.)

The comb of the Wyandotte is rose and of medium size. It should not be smooth, but show slight indentations or be pebbled, with a texture somewhat similar to the wattles. It should be curved, following the shape of the head, and should terminate in a well-defined spike which follows the neck. See Fig. 2, head 4. The spike should not, however, press against the neck so closely as to cause a depres-

sion in the feathers. If the comb shows any raise it should
be at the front and not at the back of the head. Frequently
the comb is too large and beefy, or too high. See Fig. 39.
This is especially true of males. Sometimes the spike is
embedded in the rear of the comb, as though it had been
driven into it. Occasionally there is a double spike to the
comb. A spike extending straight out or inclining upward
is a defect.

The principal defect found in the eyes is the occurrence
of those too light or green in color. This may be due in
some cases to age, and, if that is the case, should not count
as severely against an older bird as a breeder as though it
existed in the bird as a cockerel or pullet. A sunken eye is
also undesirable, and this appearance may be given to the
eye by overhanging brows.

The ear lobe should, of course, be red, the principal
defects being paleness and white in the lobe. (See page 71.)

The head and comb should be round, as well as other
sections, the curve starting from the end of the bill. A
narrow head should be avoided. The neck should be short
and well curved, well furnished with abundant hackle. If
the neck is too long, the curve is not so good, and it tends to
break the curve of the round bird. The neck should not
show a flatness in the hackle at the base of the skull, but
this place should be well filled in and rounded out so as to
carry out the unbroken line of the head and neck.

The back gives the appearance of being short. As a
matter of fact, it should not be extremely short, nor should
it be too long, but should be medium in length. Where backs
are too short, attempts are frequently made to correct this
fault by increasing the length of the tail. This does not help
the back and only makes the tail out of proportion. Increased length of back must be obtained by an increased
length of keel. Too long a back is not, however, desired, as
it destroys the typical Wyandotte shape and tends toward
straight lines in the back. The back shows a short space

above the shoulders which is level and then rises toward the tail, blending smoothly and evenly with that section. In fact, it is difficult to see just where the back leaves off and the tail begins, and for this reason the Wyandotte often appears to have a much shorter back than is really the case. Avoid any angularity of back. See Fig. 39. The back should be broad, with a broad and well-furnished saddle in the male and a slight cushion or fullness of back held well up by a well-spread tail in the female. This gives the back line of the female, from the back to the end of the tail, a slightly convex outline without any appearance of a Cochin cushion. The breadth of back is carried out in the breadth of the body generally, so that the side line of the fowl, as viewed from above, shows smooth and even, without any hollows or indentations.

In order to be typically Wyandotte in shape, the breast must be broad, full and prominent. Too often specimens are lacking in this section. The breast should not be so low, however, as to cover the hock line, but should cut in just above the hock line. Avoid a bird showing a dish or saucer-shaped breast, that is, one showing any tendency to be concave, especially on the sides, or one whose throat shows a prominent gullet.

The wings should be carried level and should not be too long. They must also be folded snugly and held in place. Low carried, slanting wings are more common in males than in females. The wing points should be well covered, so as to give a smooth, unbroken line to the side of the body, with no indication of any indentation. Well-covered wing points are more likely to be lacking in the female. A bird with a slipped or split wing should be discarded as a breeder.

The tail should be fairly short and well spread, especially at the base. If the tail is too high and long, it gives the appearance of a short back. A pinched tail is to be avoided. The sickle feathers of the male should be pliable and of medium length, so that they just nicely curve over the ends of the main tail feathers, giving the tail a short, cobby appearance. See Fig. 40. While the carriage of the tail is somewhat higher than that of the Plymouth Rock, especially in the female, it should not be too high. The top of the tail should be about on a level with the junction of the head and neck. A fairly f u l l fluff is desired, but this should not be so full as to h i d e the thighs. The legs, a s mentioned before, should be strong, well spread, and the shanks and toes o f m e d i u m l e n g t h. The shank is shorter than that of the Plymouth Rock, with the result

Fig. 40. White Wyandotte male with tail feathers so long as to destroy the short, cobby appearance of tail desired. (Photograph from the Bureau of Animal Industry, United States Department of Agriculture.)

that the fowl is set closer to the ground, just revealing the hocks, but should not be so short that the hocks are not shown. The shank should be round and full, tapering from the hock to the feet. A flat leg or shin, sometimes called "hawk leg," must be avoided. Stubs and down occur rather frequently and the former in particular must be rigidly selected against. In color the shanks and toes should be a good bright yellow. Some difficulty is experienced in the legs showing a green or willow color or green spots. Also, in the Partridge Wyandottes the shanks of the females in particular show dark to a marked degree.

The feathering of the Wyandottes is somewhat looser and more fluffy than that of the other breeds of the American Class. To maintain this character, feathers of too coarse a texture must be avoided. However, the other extreme should not be gone to and the strength of the web should be maintained or there is danger that the feathers will become so fluffy as to be Cochiny in appearance.

In mating Wyandottes the following defects common to the breed must be guarded against in so far as possible: too large and coarse birds; too small birds; too rangy or too narrow-bodied birds; lack of breast; too large, too beefy, and too high comb; spike of comb turning up or straight back; a raise in the back of the comb; spike embedded in rear of comb; double spike; too light or green eye; sunken eye; white in ear lobe; narrow head; too long a neck, not arched; a flatness in the hackle at base of skull; too short or too long back; angle between back and tail; dish or saucer-shaped breast; prominent gullet; slipped or split wing; too long tail; pinched tail; too high tail; shanks too long or too short; flat or "hawk" leg or shin; stubs and down; willow or green shank or green spots on shanks; feathers lacking strength of web so as to appear Cochiny; back narrow across saddle and at base of tail so that it fails to carry the width across the shoulders for the entire length of the body; too low on legs so as to hide hocks; fluff so full as to hide hocks.

The Silver Wyandotte

In mating this variety it is customary to use a single or standard mating. While some breeders claim that double mating is necessary, high class exhibition specimens of both sexes have been and are being produced from single matings. Birds of both sexes which are as near standard as possible should be selected. The lacing wants to be narrow and as clean and clearly defined as possible, the line between the black and white being very distinct. Be sure that the black lacing does not show a white edging, which is known as

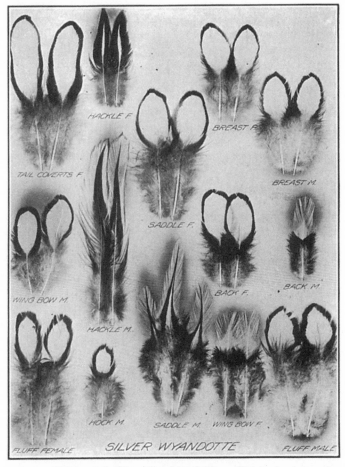

Fig. 41. Well marked Silver Wyandotte feathers. M indicates male and F female. (Photograph from the Bureau of Animal Industry, United States Department of Agriculture.)

frosting, as this is very objectionable. This frosting is especially likely to occur in females in all sections. A female desirable for breeding in other respects, but showing a little frosting in breast, should not be discarded, but offset this weakness by mating her to a male free from frosting in breast. In males frosting most frequently appears in breast and fluff. Note the frosted breast and fluff feathers of the male in Fig. 41.

Fig. 42. Standard or exhibition colored Silver Wyandotte male. Compare the feather markings with those of the male shown in Fig. 43. (Photograph from the Bureau of Animal Industry, United States Department of Agriculture.)

It is also important that the lacing be narrow so that the white centers are large. However, the extreme narrow lacing of the Sebright is not desired, as a heavier lacing is more attractive in a bird of this size. The narrower the lacing, the more open it is said to be. The center of the feather should be oval in shape, following the shape of the feather, and should show no tendency to be pointed. It is also very important that the centers of the feathers are clean white and free from any black streaks, specks or penciled markings. Where this color occurs in the white, the feather is said to be mossy. Often females molt mossy after the first year. Those that have held their clean lacing are more desirable as breeders, and the male should also be out of a hen that has molted absolutely clear on back. It is desirable that the male be well laced in saddle, back, fluff and wings especially. It is also important that the feathers of the wing bars of the male have clear white centers and be free from smuttiness, that is, the black of the lacing extending into the white center. In both sexes the lacing should be as even as possible in all sections, so as to avoid any appearance of patchiness. The birds for breeding should show good black primary feathers in the wing, with only a lower edging of white. There is, however, a tendency for the primaries to

Fig. 43. Hen feathered Silver Wyandotte male. Males showing this character of feathering and markings which closely resemble that of the female, sometimes occur and are valuable to use in heading a pullet mating where double mating is employed. Compare with Fig. 42. (Photograph from the Bureau of Animal Industry, United States Department of Agriculture.)

show too much white, but this is not a serious breeding defect.

It must be kept in mind that while most of the feathers of both sexes are white laced with black, the hackle of the female and the hackle and saddle of the male are black laced with white and with a white center extending along the shaft. It is important that the white lacing extend clear around the point of the hackle feathers. If it does not, the black ends of the feathers will cause a dark appearance or a dark ring where the neck joins the body. It is hard to get good laced hackles in the females, as they are inclined to be black with little or no lacing and striped with white along the quill. A slate under color is desirable in both sexes. The males, however, are particularly likely to be weak in this respect, showing a white under color next to the skin.

In breeding this variety for excellence of lacing there has been a tendency to overlook shape, so that the birds are likely to be lacking in type. It is most important to give type serious consideration in selecting the breeders.

In mating this variety, guard against the following defects, in so far as possible, in addition to those common to the breed (page 102) : smutty wing bars and shoulders; too narrow or too heavy lacing; frosting; mossiness, both in young birds and in hens that have molted mossy; lacing of hackle not extending around the end of the feathers.

The Golden Wyandotte

In this variety the males have a tendency to be a trifle too rangy and too narrow, and in consequence to be a little under weight. The females are likely to lack a trifle in breast and not to be quite broad enough. In breeding, the single or standard mating and the double mating systems are both employed. The single mating seems to be preferred.

In the single mating, both the sexes should be as near standard as possible. The mating is practically the same as that of the Silver Wyandotte, except that a golden bay color replaces the white of the latter. Select for breeders birds showing clean, narrow, distinct lacing, with good, large golden bay centers to the feathers. Avoid a golden bay outer edging of the black lacing which is called frosting by some breeders, and which corresponds to the frosting in the Silver variety. This is likely to occur in all sections of the females and in the breast and fluff of males. Avoid mossy feathers and females as breeders which have molted in mossy; that is, where there is a tendency for the black to mix with the bay. Use a male that is out of a female which has molted absolutely clear on back. He should also be open laced on breast and fluff.

As in the Silver Wyandotte, the lacing is reversed in the hackle of female and the hackle and saddle of male, showing a golden bay lacing instead of a black lacing, while the feather is black with a golden bay center extending along the quill. See that the lacing of the hackle feathers extends clear around the ends of the feathers, thus preventing them from being black tipped. If males have good, clear striping in the hackle and saddle, the hackle will usually come good in the female offspring.

In selecting for color, do not lean toward the yellow. The correct color is golden bay, which should approach a mahogany shade. Too dark a shade must not be selected, however, or the females will prove troublesome by molting in mossy.

In the double mating the breeders should be selected as follows:

Cockerel mating.—Use a standard or exhibition male which has a good green sheen in all black sections and is free from purple, and whose under color is very dark slate. It is important that the male in this mating have a good comb. The females should be a little darker than standard, the lacing being even in all sections, but somewhat heavier than in the exhibition female, but as distinct as possible. If weak combs have to be used on either sex in this mating, it should be on the females and not on the male.

Pullet mating.—The male for this mating should be a trifle lighter in color than the standard, the hackle striping and lacing should be very distinct, but the shade of color should be lighter. The under color likewise should be a lighter shade of slate. The breast lacing should be very clear and distinct and standard in color. The feathers of the back should have good, large golden centers, running larger in this respect than those of the cockerel-bred or exhibition male. If the lacing is clear and black, it will do no harm if the golden centers of the feathers show a little moss. If males are used having a fluff powdered with golden, they tend to get females with powdered fluff and a good deal of frosting, which are not desired. The females used should be of good size; that is, up to standard or a little larger. In other respects they should be standard. Females showing a small golden center in hackle feathers tend to get females with good clear centers in back. It is important that the combs of the females be small and as near perfect as possible, any defect in comb in this mating being in the male rather than in the female.

In mating the Golden Wyandotte, the following defects, in addition to those common to the breed (page 102), must be avoided in so far as possible: too rangy or too narrow-bodied males; males too light in weight; females lacking in breast; females not quite broad enough; smutty wing bars and shoulders; golden bay outer edging to the black lacing, sometimes called frosting; mossiness, both in pullets and in hens after they have molted; lacing of hackle not extending around the ends of the feathers; males showing purple barring in black sections.

The White Wyandotte

In uniformity of type and in excellence of type in general, this variety is undoubtedly superior to the other varieties. In breeding, only the single or standard mating is used.

Both the male and the females should be as near standard as possible. However, a female a trifle lower than standard, or one which does not show the hocks and which is therefore too low for exhibition, should not be discarded on this account from the breeding pen, as there is a tendency for the offspring to come too high. In the opinion of some breeders, the female transmits type better than the male, and they believe that a perfect type Wyandotte cannot be produced from a poor type female. She should be, therefore, the guide for type, while if it is necessary to use a weakness in color, this should be in the female and the male should be free from it. In color both sexes should be as pure a white as possible, free from any creaminess, brassiness, black ticking or any black or foreign color in the quills. If ticking occurs in the flock it is important that the male used be free. Brassiness is not troublesome, as this variety is now very free from this defect. Creaminess in the under color is also not particularly troublesome, although quite frequently encountered.

In breeding this variety the following defects, in addition to those common to the breed (page 102), must be guarded against in so far as possible: ticking in plumage, especially in main wing and tail feathers; black or foreign color in quill; creaminess of under color; duck feet, that is, rear toe turning forward; green spots on shanks or green or willow shanks; solid black in wing and tail.

The Buff Wyandotte

This variety has a tendency to be somewhat more rangy and not quite as smooth and good in type as some of the other varieties.

In mating it is unnecessary to resort to double mating, as high class standard specimens of both sexes can be produced from a single or standard mating. Double mating, as in other buff breeds, is, however, sometimes employed. In mating this variety exactly the same considerations of color apply as in the Buff Plymouth Rock, and matings should be selected upon the same basis. Willow or greenish legs are somewhat more prevalent in Buff Wyandottes than in other buff varieties, and must therefore be especially guarded against. For other defects to guard against, see the general description for the Wyandotte (page 102), and the color defects as described for the Buff Plymouth Rock (page 89).

The Black Wyandotte

This variety runs very good in type, but has somewhat of a tendency for the fowls to come too small, especially the females. While the standard calls for yellow or dusky yellow shanks, this is very difficult to get, and breeders as a rule are not so particular if the shanks show some black. It is, however, desired to have a nice yellow color to the bottom of the feet, which is also hard to get. In breeding this variety a standard mating is used. Both the male and females are selected which approach the standard as nearly as possible both in type and color. The plumage should be a good black with a greenish sheen, and should be free from any foreign color. Foreign color is most likely to occur as gray in the hackle and saddle of the male, or as gray in the wings of both sexes. Sometimes the males come with a red or straw striping in the hackle, and occasionally with a silver striping. Less frequently this striping occurs in the saddle. Males showing any of these colors should be avoided in breeding. A purple barring is also very troublesome in the Black Wyandotte and must be carefully selected against in the mating. The comb tends to run very good. In mating this breed the following defects, aside from those common to the breed (page 102), should be guarded against in so far as possible: the bottoms of the feet not a good yellow; gray in hackle and saddle of male; gray in wings of both sexes; red, straw or silver in hackle of male; purple barring; stubs; and too small size, especially in females.

The Partridge Wyandotte

In breeding this variety, it is more usual to employ only the single or standard mating, although the double mating system is sometimes used. Only the single mating is given here.

Select birds of both sexes which are as near standard as possible. The male should be a rich, bright red in all the red sections. Avoid those birds which run too dark in the red sections and also those which run to a light or lemon-colored hackle. Be sure that the red edging of the hackle and saddle extends clear around the end of the feathers, as otherwise the black tips of the feathers will cause a smutty appearance and form a black cape or ring at the base of the hackle and end of the saddle. Be sure that the shafts of both hackle and saddle feathers are black. If they are yellow, it is apt to throw shafting in all sections of the females, particularly in the breasts. In a single or standard mating it is better not to use a male showing any red in breast, as it is possible to get males with solid black breasts and yet have some red mixture or tinge in fluff and thigh. Some of the best males in other respects from a single mating will have the black in wing bows laced with red. The males should show lots of luster in the red sections, while the black should have a greenish sheen free from purple barring. It is of the utmost importance that all the red sections shall be even in color. The under color should be slate, but there is a tendency for the under color of hackle and saddle to be white. It is best to breed a male without this white in under color, but a bird of outstanding surface color may often be used, even though showing white in under color of hackle and saddle.

In selecting the females for the matings, use those showing a ground color of mahogany, being careful not to get this color too dark, as the darker birds have a tendency to run too dark in hackle. There is also a tendency for the females to run to a lemon or orange color, which is too light and should be selected against. There should be no metallic sheen apparent. There is a tendency for the penciling to run too dark, that is, the black penciling running wider than the red, when it should be the same width. Select females with broad, open penciling rather than too fine, narrow penciling. If the latter type of penciling is selected, there is a tendency for the penciling to be lost entirely in the progeny, the feathers coming stippled instead of penciled. Females are apt to be weakest in penciling in the back and fluff, where they often show stippling, barring or broken penciling. Females are rarely as good in penciling as pullets as they are after they have molted in as hens. Therefore, do not be too hasty in disposing of a pullet which shows promise, as she may molt into a splendidly penciled hen. Avoid females showing shafting, which is most likely to appear in the breast. In under color the females run very good. In leg they usually show a dark color, so much so, in fact, as to make it possible to tell the sex of day-old chicks by this means with reasonable success, as the males usually come with good, clear yellow legs. It is well to select females showing as good yellow shanks as possible, as they are desired.

In breeding this variety, the following defects, in addition to those common to the breed (page 102), should be guarded against in so far as possible: too dark red in males; lemon hackle in males; failure of the red edging of male hackle and saddle to extend around the end of the feathers; yellow shafts in hackle and saddle of male; red in breast of male; purple in the black color of males; white in under color of male hackle and saddle; too dark red color of females, especially in hackle; too light a red in females, running to lemon or orange; metallic sheen in females; too dark penciling, that is, the black penciling wider than the red of the feather; too fine or narrow penciling; stippling, barring or broken penciling in back and fluff; shafting in breast of females; dark color in legs of females; stubs in both sexes.

For additional information in regard to breeding partridge color, the reader is referred to the Partridge Cochin (page 147) and to the Partridge Plymouth Rock (page 92). with special reference to the double mating sometimes employed.

The Silver Penciled Wyandotte

In breeding this variety both single and double matings are used.

Cockerel mating.—Select a male standard both in type and color. Be sure that both hackle and saddle feathers have a clear, sharp stripe of pronounced black, that the breast is black, free from any white frosting or ticking and that the wing is black and white, free from any gray in flight feathers.

Select females of good type but of dark color. The penciling should be black and the general color of some of the feathers may be so dark that the penciling is indistinct. The hackle feathers should be sharply and distinctly striped. The flights should be dark or black, showing no gray.

Pullet mating.—Use a rather light-colored male. His breast, thighs and fluff should show some frosting and white ticking or splashing. A male should not be used, however, with too much light color or mottling of white up under the throat, as the females from him are apt to run too light in that section. An occasional male will even show some lacing of gray on the sides of the body, and these are especially valuable pullet breeders. The saddle feathers should show as distinct a stripe as possible, but a little penciling in the saddle striping is desirable in a pullet breeder. The hackle feathers should have a strong black stripe in order to offset a tendency toward too much penciling in the hackles of the female offspring.

The females for this mating should be standard in color and type.

A single or standard mating may be used with considerable success. In this mating, both the male and female should be as near standard as possible, both in type and color.

In mating this variety, the following defects, in addition to those common to the breed (page 102), must be guarded against in so far as possible: red on wings of males; solid white in flight and main wing feathers of males other than the white edging of the primaries and secondaries called for by the standard; brassy backs in males; inclination to a brownish cast instead of gray in females, which is apt to increase with age; gray in flight feathers.

For additional information on breeding Silver Penciled color, the reader is referred to the Dark Brahma (page 140), whose color scheme is identical with that of the Silver Penciled Wyandotte.

The Columbian Wyandotte

In this variety the black is often inclined to be faded instead of intense, and it fails to show a strong contrast to the white. To offset this tendency, breeders with a dark slate under color are used, those being preferred which have a clear white surface color and a half inch to an inch of bluish slate under color on the feathers, running to white next to the body.

Both the single or standard and the double mating systems are used in breeding this variety. The single mating is more common and is advised.

For the single mating, select both males and females of standard color and type as near as possible, whose hackle feathers are a clear, greenish black with a white edge running all the way around the ends of the feathers. The black and white should be positive and distinct. The tail should be a solid greenish black. The tail coverts of the female and the lesser tail coverts of the male should be greenish black with a narrow white lacing. The saddle feathers, especially the saddle hangers, have a little narrow black stripe at the ends. The tail coverts of the male should be distinctly and narrowly laced. the lacing extending the full length of the feathers and being free from any brown color. In surface color, both sexes should be a clean, distinct white with a bluish slate under color, or even quite a dark under color, even to the extent of the fluff showing a bluish tinge on the surface. It used to be common to use females showing black on the surface of the back in order to get black wing feathers. Such females are not now in favor as breeders, nor are males showing too much striping in the saddle. To attain the greatest success, keep away from breeders with pure white under color and save as breeders those showing no black in the surface of white sections. It is also necessary to guard against brassiness, as this is a serious defect. Discard any brassy birds as breeders. Brassiness is more apparent in males and may appear on hackle, back, wing bows and saddle and saddle hangers. In females brassiness is most likely to appear in the white lacing of the hackle.

Cockerel Mating.—In using the double mating system, the cockerel mating is as follows: a standard colored male with strong green sheen in all black sections is desired. The hackle striping should be clear and distinct and carried all the way to the skin, while the saddle is not heavily laced and the striping runs only about one-quarter way down the feather with a distinct break and then bluish slate under color. The back should be absolutely clean in color and the main tail a solid black running clear to the base of the feathers. The fluff should be white with bluish slate under color, ending with white at the skin.

The females should be of standard type. The feathers should be broad. The hackle lacing should come well around in front of the neck, while the back should show no ticking in surface. The under color of back should be a strong bluish white. The wings should show strong color, the black and white being distinct and clearly defined. Good, narrow, sharp lacing is desired in the tail coverts. In general, the females for this mating should show darker color tones than the standard female.

Pullet mating.—Select a male of good type and head points, but with a trifle lighter color in all sections, including under color, than is standard. Both the hackle and saddle feathers should be long and flowing, and very slightly striped. The tail should be profusely feathered with coverts which are distinctly and narrowly laced, the lacing extending the full length of the feathers and being free from any brown color. Strong black coloring in the wing primaries and secondaries is very important, the black and white being distinct and free from peppering.

The females of this mating should be standard, both in type and color. A slate under color of fluff is desired. The lacing of the hackle and the tail coverts should be as distinct and well defined as possible.

The following defects, in addition to those common to the breed (page 102), should be guarded against, in so far as possible, in breeding this variety: stubs and down; too great a length of back; birds with too light or faded appearing markings; black in surface of back of females; pure white under color; the white lacing not extending clear around the ends of the hackle feathers in both sexes.

Additional information as to breeding Columbian color will be found under the Light Brahma (page 134), the color scheme of which is identical with that of the Columbian varieties. It must be kept in mind, however, that the Columbian Wyandotte is a newer variety than the Light Brahma, and that it has not as yet been developed to the same degree of perfection.

The Java

The type of the Java is quite distinctive. It is a good-sized bird and has a body rather rectangular in outline, with good length both of back and keel. In fact, this breed is longest in body of any of the American breeds. It is set on legs about like the Plymouth Rock, but not so low as the Wyandotte. It has good breadth of back and body and well-rounded, prominent breast. In general type it is much like the Rhode Island Red, but is larger and has a longer tail, carried more erect. The back line and the body have a slight downward slope from front to rear, instead of being level like the back and body of the Rhode Island Red. The carriage of tail is higher, and tail and back join in a gradual slope.

There is a tendency in this breed for the fowls to lack size, especially in the females, and along with this lack in size goes a lack in the size of bone and a shortness of back. There is also somewhat of a tendency toward stiltiness; that is, the legs and neck are too long. In selecting breeders, attention must be given to keeping up the size of body and bone and the length of back, while the tendency toward stiltiness must be offset by the selection of breeders with shorter necks and legs.

The comb, which is single and upright, should be a little larger than that of the Plymouth Rock. Some of the combs come too large and in the females many are inclined to lop. Too small combs should not be selected, for this is considered by some breeders to be associated with poorer egg production. There is an inclination toward side sprigs, especially in the Mottled variety.

The ear lobes should be red, and any tendency toward white must be guarded against in selecting the breeders; but here, as in the Plymouth Rocks (page 71), paleness or lack of color due to poor condition must not be considered as serious a breeding defect as positive white. This is more troublesome in the Mottled than in the Black variety.

The eye color is not the same for the two varieties, being darker or more on the black in the Blacks and lighter or more on the red in the Mottled. In both, however, there is a tendency for the eyes to come too light, especially in the Mottled, which sometimes shows a pearl eye.

The shanks should be clean, but in both varieties the occurrence of stubs is not uncommon.

In breeding Javas, it is therefore necessary to guard against the following defects in so far as possible: too small size, especially in females; back not long enough; bone too fine; stiltiness, that is, legs and neck too long; white in ear lobes; too large combs; lopped combs in females; side sprigs; stubs; light eyes.

The Black Java

In breeding this variety, only the single or standard mating is used.

The mating is essentially like that of other black matings. The birds of both sexes should be as near standard as possible, using birds with pronounced green sheen but without any purple. Although this variety is quite free from purple barring, more so, in fact, than most black fowls, any tendency in this direction must be carefully guarded against. The color of both sexes should, of course, be black throughout, free from any foreign color, which is most apt to occur as red or straw in hackle, back, wing bows and saddle of males, or as white or gray in the wings and at the root of tail. However, this white is not very prevalent and consequently not very troublesome. The under color should also be black. There is somewhat of a tendency for the under color to lighten in hackle of male and somewhat less so in back and saddle of male. Black legs are preferred if they can be found, combined with yellow bottoms to the feet, but where legs and toes are black the feet are apt to be white or pink on the bottom. Where the legs tend toward a willow or green color, the bottoms of the feet are more likely to be yellow, and there is less trouble with the bottoms of the feet coming white or pink.

In breeding this variety the following defects, in addition to those common to the breed (page 118), must be guarded against in so far as possible: purple barring; red or straw in hackle, back, wing bow and saddle of males; white or gray in wings and at root of tail; white or light under color in hackle, back and saddle of males; pink or white bottoms to the feet.

The Mottled Java

As with the Black Java, only the single or standard mating is used in breeding this variety.

Breeders of both sexes should be selected which are as near standard as possible. In this variety there is a tendency for too many light-colored birds. There is also a tendency, especially in the females, to show more white with each successive molt. Therefore the darker colored birds, both males and females, should be selected as breeders.

The white tips to the feathers should be small and distinct, so that they afford a good contrast to the black of the feathers. The line between the black and the white should be as distinctly and sharply defined as possible, so that the white is free from any black or gray color.

There is a tendency for birds of both sexes, but more to a slight extent in males, to show too many white feathers in the wing bows. Males are also apt to show red or brass in the hackle, back, wing bow and saddle. There is a tendency toward too much white or toward solid white feathers in the tail in both sexes. Sometimes males come with a solid white tail. Such birds should not be used as breeders.

The shank should not be too dark in color, but should show some yellow. It is not desirable to select breeders with green or olive legs. There is not as much trouble in this variety as in the Blacks with white or pink bottoms to the feet instead of yellow.

In mating this variety the following defects, in addition

Fig. 44. Well marked Mottled Java feathers. M indicates male and F female. (Photograph from the Bureau of Animal Industry, United States Department of Agriculture.)

to those common to the breed (page 118), should be guarded against in so far as possible: birds too light in color; white tips too large; indistinct white tips showing some black or gray; too many white feathers in wing bows of both males and females; red or brass in hackle, back, wing bow and saddle of males; green or olive shanks; white in tail; white, pinkish or flesh-colored bottoms of feet.

The Dominique

The Dominique is in type much on the Leghorn or Hamburg order, but larger in size. The carriage is upright, with a general appearance of alertness. The tail is carried slightly higher than in the Leghorn or Hamburg. It is well spread and in the male is furnished with long, curved, sweeping sickles. The comb, which is rose, should be practically the same as that of the Hamburg, although larger in proportion to the size of the fowl. The spike, like the Hamburg, turns up slightly at the end.

In color the two sexes are different. The female has a color scheme of dark bluish slate approaching black, and very light slate approaching white, arranged in alternate, irregular patches or bars across the feather. The light and dark markings are about equal in width. The under color is slate, with indistinct barring. The male is one or two shades lighter in color than the female, this being caused by the fact that the light markings are wider than the dark. The markings as a whole are narrower in all sections than in the females.

In mating this variety, only the single or standard mating is used. While the tendency is the same here as in Barred Plymouth Rocks, for the males to come lighter than the females, the single mating is possible because the standard calls for a male showing this lighter shade of color.

For the mating, select medium-colored females, with rich yellow legs and red eyes. Discard the females which have a tendency toward more of the dark slate than the light slate markings, as they are too dark in color and also nearly always show black on the front of the legs. Do not use a female which has not a well-spread tail.

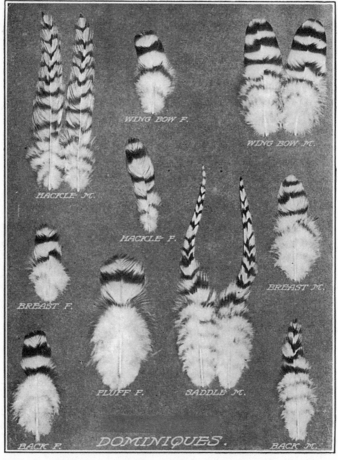

Fig. 45. Well marked Dominique feathers. Contrast the markings with those of the Barred Plymouth Rock as shown in Figs. 32 to 35. M indicates male and F female. (Photograph from the Bureau of Animal Industry, United States Department of Agriculture.)

Select a male one or two shades lighter than the females. He should have upright carriage, long curving sickles carried well out, clean legs and red eyes.

Care should be taken to see that the comb and head are good in both sexes. It must also be remembered that the breed is of medium size; therefore birds of both sexes should not be over standard weight.

The following defects must be guarded against, in so far as possible, in mating this breed: pinched tail; too large comb; comb not straight on head; comb with hollow center or hollow along sides; too dark color in females; light eyes; dark or black on legs of females; stubs; shafting; brownish tinge or metallic sheen to plumage.

The Single Comb Rhode Island Red

In size this breed is medium, being a little smaller than the Plymouth Rock. Birds of about standard weight should be selected as breeders. Those larger than standard are apt to be poorer layers and breeders, while birds much smaller than standard should not be used as there is somewhat of a tendency for the individuals of this breed to run too small in size. In type it is quite distinct, approaching most nearly that of the Java, than which, however, it is smaller. The body of the Rhode Island Red should be carried perfectly level or horizontal, and should be long, approaching in shape as seen from the side as nearly as possible to an oblong or rectangle. The lines of back and keel should both be level and parallel to one another. The base line of the wing should be parallel to the lines of back and keel and the wing should have no tendency to drop down, as this is a serious defect. The back should not only be flat from front to rear, but should also be quite flat from side to side, showing no tendency to slope from the backbone to the side, and should therefore show no indication of a ridge along the backbone. In this respect the Rhode Island Red back is flatter than that of the Java. The breast must be prominent in order to fill out the rectangular shape. A line dropped through the base of the beak should just clear the front of the breast. The body and back should be broad, and this breadth should be carried out the full length of back. The tail should be well spread and low carried, that is, not higher than standard, in both sexes, but should not be drooping.

The bird should be well balanced, that is, the legs should be under the center of the fowl. The shanks should be fairly stout and of medium length, but not so long as to give the bird a stilty appearance. The neck also should be medium in length, as a long neck also tends to give a bird a stilty or too rangy appearance. Shanks should be a good yellow or reddish horn color, the red extending to the end of the toes, as this indicates strong breeding. Legs exceedingly yellow in color are apt to be associated with too light a color of plumage. The shanks and toes should be free from stubs and down, which are serious faults in this breed.

The comb should be a good, clean-cut, evenly-serrated, 5-point single comb of fine texture, of medium size and in proportion to the bird. However, the comb tends to run rather large and also rather irregular or wavy in shape with uneven serrations. A comb which is a little too heavy, if clean, of fine texture and evenly serrated, is preferable to the irregular combs, as these are very troublesome in breeding. There is a noticeable tendency toward side sprigs, which must be guarded against. See Fig. 12.

A red eye is very desirable in breeders of both sexes. However, a male good in other respects should not be discarded as a breeder if his eye is bright and on the bay order. The color of the eye of females tends to fade with laying, so that many hens show light or even green eyes, and it is rather difficult to find good eyes in old hens. However, hens with bright eyes on the bay order were probably strong in eye as pullets and can safely be used as breeders.

The beak should be reddish horn color. Sometimes black or dark, usually as a streak, is present in the beak. This is not desired and should be selected against. White in the ear lobes occasionally shows and while birds with this defect should not be bred if it can be avoided, it is not very common or very troublesome in this breed.

In mating the Rhode Island Red, a double mating is often employed especially for the sake of color. However, this is unnecessary if a proper selection of breeders is made, as good specimens of both sexes can be produced from the same mating. To avoid the double mating, do not attempt to use the extremely dark males, as the females from them are often poor in color, tending to be mottled. The mating should consist of a rich, snappy colored male of even shade in hackle, wing bows, back and saddle, but not of extremely dark color, and females which are dark, rich and even in color. Individuals should not be selected for mere darkness or depth of color. In addition to depth of color, the plumage should be a lustrous, bright and live color which is a decided red, and not a flat, dead color which shows as brown or chocolate. From such a male, the pullets produced will be much more uniform and a greater percentage of the rich, soft, even color so much admired and so much desired. From such a mating fine stock of both sexes can be produced.

There is a decided tendency for the birds to show unevenness in surface color. In males, the hackle and saddle are likely to be lighter than the back and wing bows, and in females the hackle is often a lighter color than the rest of the surface. Birds of even color are especially desired and are especially valuable as breeders.

The females for the mating should not be too strong

in the black markings, that is, they should not show too much black in hackle and wings. In fact, a hackle free from black ticking and wings in which the black markings are faint are preferable in females, as the male with strong black markings in wing will give about the right amount of ticking in the pullets. If the black markings are strong in these sections of the females and also in the wing of the male there will be a tendency to produce black lacing in the hackles of the male offspring. The male used in the mating should have a standard black wing marking unless it becomes necessary to use females which are strong in the black of wing and hackle, in which case the male should be weak in the black of wing markings, but extra strong in rich, deep red pigment.

Fig. 46. Single Comb Rhode Island Red hen showing split wing folded. (Photograph from the Bureau of Animal Industry, United States Department of Agriculture.)

Smut is apt to occur in under color of both sexes. Never use a smutty bird if it can be avoided. If it becomes desirable or necessary to use a female with very dark surface color, and fairly clear under color, but showing just a tinge of smut and with considerable black in wing, be sure that she is mated with a male absolutely free from smut and not too heavy in black points. If a male showing smut is mated with heavy colored females, it is apt to show as black in the surface of the pullets produced and in black-laced hackles in the cockerels.

The under color should be as deep a red as possible. There is a tendency for it to run too light and in the hackle, back and saddle of males especially it may even be white. This white often develops with age, so that its occurrence in a male which was sound as a cockerel is not such a serious defect.

Rhode Island Reds, especially females, usually fade in color up to the time of molting. This is in consequence

Fig. 47. Same hen as shown in Fig. 46 with wing spread to show the split in wing. (Photograph from the Bureau of Animal Industry, United States Department of Agriculture.)

a poor time of year to judge as to their color quality. The color of the plumage of hens is generally better after the molt than just before, but this is not always true, and few hens ever approximate their pullet color. A good deal depends in this particular upon the time and manner of molting and upon the condition of the bird. A hen exhausted by heavy laying is not apt to molt in so well as one that has not been laying so heavily. A hen which molts gradually is apt to show a mottled color when the molt is completed, due to the fading of the first feathers grown by the time the last are in. Occasionally hens show a considerable number of white tips to the hackle feathers after the molt and often they will show a considerable amount of black in the surface, even though they were free from this color before they molted. Consequently, a knowledge of the pullet form of hens with respect to color is important for intelligent mating, and it is extremely valuable to make notes on the young stock as a guide for future breeding. Hens which return to good color after the molt, approaching as closely as possible their color as pullets, are preferable as breeders.

It is impossible to judge the quality of stock as to color from the color of the day-old chicks. Often the best colored stock will show a very light colored down as chicks.

In mating this variety, the following defects must be guarded against in so far as possible: Black in beak; too large or too small birds; males with too long and too thin shanks; white in ear lobes; too large and irregular combs; uneven serrations and points of irregular lengths in combs; side sprigs; too light eyes; too much ticking in hackle of females; narrow backs and pinched tails in both sexes; stubs and down; legs too yellow and lacking in horn color; deficient breasts, especially in males; smut in under color; white in under color, especially in hackle, back and saddle of males; black in surface, especially on shoulders of both sexes and in hackle of males; uneven shade of color in both sexes; slipped or split wings.

The Rose Comb Rhode Island Red

This variety is identical in color and type with the Single Comb Rhode Island Red. In mating, therefore, exactly the same principles are to be observed except in the matter of comb, which is rose. It is necessary to guard against combs in both sexes which are too large or too high, in which the spike does not tend to follow the neck, and which show hollows either along the side or in the center of the top.

The Buckeye

This breed is one of the newer American breeds and is not very common. While the standard calls for a bird which resembles in appearance most nearly the Cornish, as a matter of fact the birds as found seem to be more on the Rhode Island Red type. They differ from the Rhode Island Red in not being quite so rectangular in outline, and in not having quite as deep or low-carried a breast, while the body slopes slightly from the front to the rear. There is a tendency for the birds of this breed to come too shallow in body. This should be guarded against, as birds deep in body are desired. Any tendency toward the Wyandotte shape should be avoided, as the body is more oblong than round. The Buckeye is the only breed of the American Class which has a pea comb. The shape of comb runs very good, in fact, this is one of the most perfect sections of the breed. The tail should be carried low and should be well spread. This breed is a moderately close-feathered bird, being almost like the Rhode Island Red in this respect.

In mating this breed, the single or standard mating is used. Select both males and females which are deep and long in body, and in which the body is only slightly off the horizontal. In color, select birds which are a nice, deep, rich red, avoiding birds which run to a chocolate. A smoky bar in the under color of the back is allowed in this breed.

It is necessary to select birds which are strong in wing color, as there is a tendency for a decided weakness here. The black sections of the wings are apt to be peppery instead of solid black and the same holds true to some extent in the tail. It is also necessary to avoid any black in the surface of the body. What is desired in the color of the breed is an even shade of rich, dark red color. In the male, the hackle, saddle and back should match as closely as possible. In the female the color should be even all over and as free from chocolate as possible. This chocolate color should also be avoided in the breast of the male.

In selecting the mating in Buckeyes, it is necessary to guard against the following defects in so far as possible: Too shallow body; any tendency toward Wyandotte shape; any tendency toward chocolate color in surface of female or breast of male; any tendency toward white ear lobes; stubs (this is especially likely to occur in the larger birds); peppery color in tail; black in surface; uneven shade of hackle, back and saddle of male; uneven surface color in female.

CHAPTER V

THE ASIATIC CLASS

The Brahma

The Brahma is the largest of the standard varieties of chickens. The large body, of proportionate length, breadth and depth, and well-rounded breast standing well up on legs gives the breed a majestic appearance. An excessive fluff may give a bird the appearance of greater depth than it actually possesses. In type, the Light and the Dark Brahma are identical, but in size the Light is about one pound heavier than the Dark. In order to secure the desired size in the offspring, it is preferable to use females of standard size, good frame and bone, rather than males which are standard or larger than standard, as such males are apt to be less active and less satisfactory breeders. Many birds, however, which have the desired size, frame and bone, will not, when in breeding condition, reach standard weights. In the Dark variety there is somewhat of a tendency for the birds to fail in size so that the large birds should be used as breeders. It is important that the length of body, the higher station, the more compact feathering of body and the less profuse feathering of fluff, toes and legs be maintained in order to keep the type distinct from the Cochin, which there is some tendency for the Brahma to approach. In general, the birds used as breeders should be rather oblong in shape, and those tending to be heart-shaped or those square-shaped should be avoided.

The head is distinctive. The head should be wide, the skull projecting over the eyes. There is somewhat of a tendency for the heads of females to be narrow or snaky in appearance and this must be avoided. The color of eye tends to come good in both varieties. The comb, which is pea, adds distinctiveness to the head. It should be small and neat and regularly formed, showing its pea character distinctly. See Fig. 2, heads 2 and 3, and Fig. 3, head 2. In the males there is a tendency for the comb to run too large or coarse. If it becomes necessary to use a male whose comb is at all coarse or irregular, it is important to offset this by using females with very small, even combs. The wattles should be rather small and almost round, but the tendency toward too small wattles, which is evident in some males, must be avoided, as it gives them a feminine appearance. The ear lobes are rather large and pendulous.

The neck should be medium in length, thick and well arched, and there should be a smooth blending of the hackle and back so that no distinct angle is formed. The shape of neck will usually be good if the length of feathers is good and in proportion. No matter how good the color in this section, do not use a male with short length of hackle feathers. The wings should be well bowed, that is, curved or sprung so that they conform to the body shape and carry out the curve of the back over the side of the body, and carried high enough to give a broad appearance to the back. The primaries and secondaries should be carried tightly folded and held in place. Avoid as breeders birds with "flat wings" or with loose or slipped wings. See Fig. 53. The broad back is carried out to a good length and is supplemented by a well-spread tail, which carries out the breadth of back. See Fig. 48. The use of females with well-spread tails and having a plentiful supply of feathers which lap well up on to the tail coverts will produce long saddle hangers in males, which are greatly desired. The tail is carried at a medium height in the male and somewhat lower in the female. The body and keel are long in proportion to the back, the breast well rounded. A sloping or shallow breast, such as is sometimes found on males, is undesirable and is especially likely to occur in cockerels. The whole body is full and rounded.

The Brahma male stands fairly well up on his legs so that he has a rather active appearance, but he should not be so high as to a p p e a r leggy. The fe-male is perhaps a trifle lower set i n proportion than the male. The best shaped cockerels are apt to be out of fe-males set a trifle l o w e r t h a n standard. Cock-erels out of fe-males which are standard in this r e s p e c t are often too leggy. The bone of the legs should be h e a v y and of good substance and should not tend to be light or small.

Fig. 48—Light Brahma male. Notice the well-spread tail and the hocks covered with short, soft feathers which in no way approach vulture hocks. (Photograph from the Bureau of Animal Industry, United States Department of Agriculture.)

W h i l e t h e shank and the outer and middle toes should be feathered, this feathering should be medium and should neither tend to be very light, in which case the middle toe is apt to be bare, nor so heavy as to approach the toe feathering of the Cochin. If it becomes necessary to breed a bird which tends to be extreme in toe feathering in either direction, this should be balanced by the toe feathering of the opposite sex. Birds with stiff toe feathering are apt to breed vulture hocks, which are more prevalent in Dark than in Light Brahmas. See Figs. 48 and 53. If, as occasionally happens, it is desirable to breed a bird, usually a female, with vulture hocks, she should be mated with a male which is scant and very soft in hock feathering. It is important, however, to avoid breeding birds which have vulture hocks, if this can possibly be done, as this defect is likely to crop out again and again.

The feathering of Brahmas, while heavy, is quite compact and should have none of the excessive fluffiness of the Cochin, which is particularly marked in fluff, front of thighs, under part of body, cushion and leg and toe feathering. The feathering should be tight or compact enough to render the birds deceptive as to weight. In breeding Brahmas, the following defects must be guarded against in so far as possible: square-shaped birds; heart-shaped birds; short backs; lack of breadth of back; too low on legs; flat or shallow breast, particularly in males; tendency toward the Cochin type; slim, narrow or snaky heads, particularly in females; too high combs and too small wattles in males; tails not well spread; too high or too leggy appearance in males; too small or too fine bone of legs and shanks; middle toe feathering absent or scanty; too heavy shank or toe feathering; vulture hocks; lack of compactness or too fluffy character of feathering; size above standard; under-sized birds; flat wings; loose or slipped wings.

The Light Brahma

In breeding this variety, the single or standard mating is commonly used. It frequently happens, however, that the best females come from one female in the mating and the best males from another, but it is seldom that the best males and females come from the same hen. Fe-

Fig. 49—Well-marked Light Brahma feathers. M indicates male and F female. (Photograph from the Bureau of Animal Industry, United States Department of Agriculture.)

males which stand well up on legs seldom produce good-shaped males, while low-set females seldom produce good-shaped females. Birds of both sexes should be as near standard, both in type and color, as possible, with special emphasis laid on certain points. However, the breeding back of the birds is even more important than the extreme excellence of their markings. The black and white of the plumage wants to be distinct in color and there should be a sharp and clean-cut line between the two colors, with no tendency for them to run into one another so as to produce indistinct or smutty markings. As there is a tendency for the lacing in the hackle of both sexes to be indistinct, this must be guarded against. The lacings should be of a silvery white color and no gray or brown cast should appear anywhere in the plumage. Use specimens which show a bluish white or even a slate under color, as such birds tend to produce a better surface color. As there is a tendency for the under color to become lighter in this breed, select birds as breeders having an under color slightly darker than that desired in the offspring.

In selecting a male, use one with a long, wide back and a well-spread tail which carries out the width of back. Be sure to keep away from the Cochin type. The breast should be particularly full and the hackle should come around well in front of the neck. As there is a tendency for the black stripe in the hackle to extend through the end of the feathers, select males in which the white edging extends clear around the end of the hackle feathers. The black stripe of the hackle feathers should be broad, the white edging narrow and the line of demarcation between the two should be distinct. If a male is used having comparatively wide white edging to the hackle feathers, the females mated to him should have extremely narrow white lacing in the hackle. The deeper the black stripe of the male's hackle runs into the under color without a break, the better will be the hackles in the offspring. A striping in the hackle feathers of the male which is of uniform width throughout nearly its entire length and terminating rather abruptly in a V-shaped point is preferable to a striping which tapers more gradually to a point, as it produces better shaped black centers in the hackle feathers of females. The male should also be laced to a certain extent on the back at the base of the tail. See Fig. 50. Males show-

Fig. 50—Light Brahma male showing good wing markings and the lacing of some of the feathers of the back at the base of the tail. (Photograph from the Bureau of Animal Industry, United States Department of Agriculture.)

ing this lacing and at the same time a bluish white or slate under color are more likely to be free from brass, while those with plain backs, that is, showing no black, and with white under color, tend toward a straw or yellow surface color, which it is important to eliminate. Brassiness in males can be largely eliminated in a few generations by breeding to females with dark under color. Males should not have too much coloring in hackle and saddle, as with this excess color is likely to go black ticking in throat, breast and fluff. Watch out for males showing purple barring in the black of tail and hackle, as this is undesirable. In order to secure the solid black tails desired in males, male birds must be selected as breeders whose main tail and sickle feathers run black clear to the skin. If it becomes necessary to use a male showing some gray in this section, it is very important to use females which have no gray or white at the base of the main tail feathers. Be sure that the primary feathers of the male are a deep distinct black and avoid any tendency toward a brownish black in this section. See Fig. 50. Gray in wings of either male or female should be avoided unless it is due to injury. White spots in the primaries of females, while not desirable, are not serious if the male to which they are mated has good black wing color with which to counteract it.

The females should also have wide backs, well-spread tails and distinctly laced hackle feathers coming well around in front of the throat. They should be well up on legs and well away from the Cochin type. In the earlier days of the breed, as in Columbian Wyandottes and Plymouth Rocks, it was difficult to get females with good strong black markings in hackle, wing and tail without black cropping out in the surface of the back. Now, however, black seldom shows in the surface, although it is found to some extent in the web of the feathers or as a black ticking. This ticking is most serious in females having good black wings, hackle and tail, and must be guarded against. It is most important never to use a male which has black in the web of the feathers of the back, while gray or brown in this section is even worse than the black and should not be tolerated. Good females can rarely be secured from such a male, as brown or black will appear in the surface of the back. Experienced breeders sometimes use dark under-colored females that have a slight amount of dark specks or ticking running up into the web of the feathers of the back in order to intensify or improve the black markings of the flock. This should only be resorted to, however, when the breeder is thoroughly familiar with the breeding tendencies of his stock, and is a dangerous practice for the amateur, as it is likely to cause an increase of the black in the surface color of the flock. This class of females, mated to a male with clean back, will produce the best laced saddles on the male offspring, while the female offspring will generally be clean on the back. Females with light under color and with black in the web of feathers of the back should not be used.

Fig. 51—Light Brahma female showing good bluish slate undercolor. (Photograph from the Bureau of Animal Industry, United States Department of Agriculture.)

Females with a good bluish slate under color are best as breeders. See Fig. 51. Attention must also be given to the lacing of the tail coverts in both sexes. Too heavy lacing in these feathers is not desired, and if necessary to use breeders of one sex showing this, it must be offset by using breeders of the opposite sex with extremely narrow lacing. The lacing here and in the hackle should be clear cut and distinct, with no tendency for the black to run into the white or the white into the black. See Fig. 50.

In mating this variety, the following defects, in addition to those common to the breed (page 134), must be guarded against in so far as possible: indistinct contrast between the black and white; indistinct lacing in hackle of both sexes; too light under color in both sexes; the white lacing of hackle feathers of both sexes not extending clear around the end of the feathers; no lacing on back at base of tail in males; brassy surface color in males; black ticking in throat, breast and fluff of male; purple barring in the black of male's tail and hackle; black or black ticking in back of females; too heavy lacing in tail coverts of females; indistinct or smutty lacing in hackle and tail coverts of females; main tail and sickle feathers not black clear to the skin; brown cast in the primary wing feathers of males; gray in wings of both sexes; white spots in the primaries of females.

The Dark Brahma

In breeding this variety it is usual to employ the double mating system. The natural drift for the person who does not know much about breeding Dark Brahmas, and who does not double mate, is toward the cockerel mating.

Cockerel mating.—The male for this mating should be standard, and as in all double-mated varieties, must be bred from a cockerel mating line. The color of eye must be good, as there is a tendency for the exhibition males to have gray eyes. Often in trying to produce males with black breasts and body, the hackle will tend to be smutty, that is, there will be a tendency for the black to run into the white edging. A bird with clear striping of hackle is desirable and should be used if possible, and if a male with cloudy or smutty hackle has to be used, this condition must be counterbalanced by the clear striping of the female hackle. There is also a tendency for the black stripe of the hackle and saddle feathers to run the entire length of the feathers, extending through the white edging at the ends of the feathers. This often occurs when there is too much black in the under color of the hackle, and, in fact, is quite likely to occur in most males that have a good dark under color and clear striping. Such males, in which the black stripe does not run through the ends of the hackle feathers, are therefore valuable breeders. There is also a tendency for the hackle to come light in under color and for this light color to run along the shaft of the feathers. The under color of the back and saddle is also inclined to run light and must be guarded against. A good slate under color is desired. There is some tendency toward brown on the

Fig. 52—Cockerel bred Dark Brahma male showing solid black breast. Contrast with pullet bred male Fig. 53. Notice the white in the toe feathers. (Photograph from the Bureau of Animal Industry, United States Department of Agriculture.)

shoulders. The breast usually comes good and black, but sometimes shows some white ticking. See Fig. 53. The fluff frequently shows white ticking or even so much white or gray as to appear frosty or grizzly. The use of dark females from a male line assists in getting rid of this defect. White sometimes occurs at the base of the main sickles and also in the leg and toe feathers, which should be avoided as much as possible. See Fig. 53. White feathers develop in the wings with age, showing frequently at two years of age. All black sections must have a good green luster, free from purple.

The females for this mating should not be the light silvery color, as seen in the exhibition female, but should be dark, although not approaching a brownish cast. It is essential that the hackle should be well and clearly striped with black and edged with silver in order to produce good hackles in the males. It is very necessary that the females be out of a straight cockerel line. The penciling need not be as clear as in the exhibition female. It is important that the females be out of a male that possesses a good saddle, as this is one of the hardest points to get in the exhibition male. It is also important that eyes and comb be good.

Pullet mating.—The male should, of course, be bred from a pullet line, and one should know the females he is out of for at least two generations back. At a distance the pullet-bred male looks a good deal like the exhibition male, but in general he is a lighter colored bird. He will not have as much black in his feathers and will be more or less mottled with white in breast and fluff. See Fig. 53. His shoulders or wing bows will also be whiter or more silvery and his wing flights and secondaries will show more light than the exhibition male. The striping is not as distinct, the saddle striping being more or less broken. Often the white or silver will form bars across the feathers, particularly in the hackle, saddle and flight feathers. This white or silver bar should be fought against as much as possible. Often the hackle is like that of the exhibition male, but if the bird is bred right, the marking in the hackle does not make much difference. Many pullet-bred males, while still in their chick plumage, show some penciling, especially in the wings, but as they gain their mature plumage this penciling is wholly or largely lost. Males showing good penciling in their chick feathers are as a rule valuable pullet breeders, but since they lose this penciling later, it is necessary to judge their quality in this respect while they still have their chick feathers.

The females for this mating should be standard, both in type and color, and, of course, bred from a pullet line. They should be of a light, soft silvery color with c l e a r, distinct concentric penciling. There is a tendency for pullets of the pullet line to be far too light in throat and upper breast. It must be remembered, however, that the females grow d a r k e r with age, so that such pullets are apt to be excellent in mark-

Fig. 53—Pullet bred Dark Brahma male showing white mottling in breast and fluff. Notice the vulture hock and the slipped wing. (Photograph from the Bureau of Animal Industry, United States Department of Agriculture.)

ings and color after they have molted in as hens. On the other hand the better penciled pullets also darken with age and are not so good as hens, tending to show too coarse a penciling in hackle. Obviously the lighter pullets should not be discarded as breeders until they are molted in as hens and a chance obtained to judge their quality. Females which are especially good in penciling of throat and breast are apt to lose in penciling of lower back and tail and vice versa. Good penciling in the fluff of exhibition or pullet-bred females is hard to get, but is very desirable. There is a tendency for a brown cast to develop with age, which is, of course, undesirable.

Some breeders advocate a single mating in this variety. Where this practice is followed, birds of both sexes should

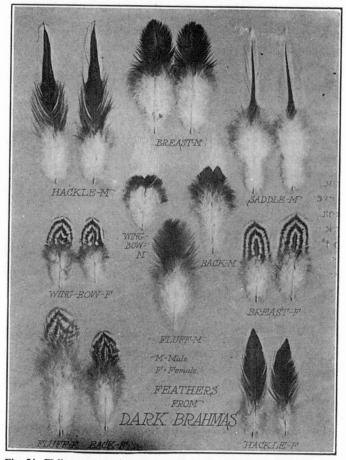

Fig. 54—Well-marked **Dark Brahma** feathers. M indicates male and F female. (Photograph from the Bureau of Animal Industry, United States Department of Agriculture.)

be selected which are as near standard as possible. It is very important, however, to know the breeding of the male, especially so far as the female line is concerned, if good pullets are to be expected from the mating. Dependence on the knowledge of the breeding in the female line is necessary because there is not much in the male's plumage by which the beginner can judge of his quality as a pullet breeder. Males for this mating may be selected with under color a little lighter than the standard calls for, as too dark an under color is likely to result in too dark ground color in his female offspring. The male may also show some frosting in lower breast and fluff, but he should be as near black as possible up under the throat, as white in that section is likely to result in females which are too light there. A male showing some penciling, especially in the wings, while in chick plumage, if he develops into a good quality exhibition male is especially valuable as a breeder for this mating, since he has the potential power to produce both males and females of good quality. See Fig. 52.

The following defects, in addition to those common to the breed (page 134), must be guarded against, in so far as possible, in mating this variety:

In exhibition males: gray eyes; cloudy or smutty hackle; hackle and saddle striping running clear through the ends of the feathers; too much black in under color of hackle; too light under color of hackle, back and saddle; brown on shoulders; white ticking in breast; frosty or grizzly fluff; white at base of main sickles; white in leg and toe feathers; white in wings, developing with age; purple barring in black sections.

In cockerel-bred females: brownish cast; indistinct hackle striping; poor eye; poor comb.

In pullet-bred males: no white in breast and fluff; white or silver bars across hackle, saddle and flight feathers.

In the pullet-bred or exhibition females: too dark general color; indistinct penciling; poor penciling of throat and breast; poor penciling of lower back and tail.

The Cochin

The Cochin is unique among breeds, largely because of its profuse and loose feathering. It is a large breed, weighing but little less than the Brahma. In type it is absolutely distinct from either the Brahma or the Langshan. It is a low set, round-bodied bird. In fact, the female with well-developed feathering is almost a ball. The breast and fluff are carried close to the ground and the breast feathering and that of the toes and of the legs in front of the hock almost completely fill in the space between the breast and the ground. In fact, the Cochins are more completely filled in and show less daylight underneath than any other of the well-known breeds. In general, the Cochin is a shorter, rounder, broader, lower and more profusely feathered bird than the Brahma. However, there is somewhat of a tendency for males to be too stilty in appearance, which is due to the fact that they are too long-legged and are not well feathered or filled in underneath. The back is very broad but very short in appearance, being even shorter than the Langshan. This appearance is quite largely due to the extreme development of the cushion, especially in the female, which is very characteristic of the breed and which gives a decided convexity to the shape of the back and tail of the female.

The comb is single and should be medium in size. In mating, select against the too large or too high combs, which are more likely to occur in males. The reddish bay eye is rather difficult to get, as there is a tendency for the eyes to run too light. It is therefore important to select breeders with good eyes if possible. The neck in both sexes is decidedly short and thick, much more so than the Brahma.

Main tail feathers which are short and soft are desired in both sexes. In the male, they should be covered with soft and profuse sickle and lesser sickle feathers. The feathering of the entire body is very profuse, loose, long and soft, more so, in fact, than any other of the well-known breeds. The only stiff feathers on the bird should be the long toe and wing feathers.

The shank and toe feathering is more profuse than in other breeds. Both the inner and middle toes should be heavily feathered clear to their extremities. The heaviness of feathering should be carried down the shank to the foot. If it tends to grow lighter toward the foot, it will cause a V-shaped junction of the shank and toe feathering, which sometimes occurs and is undesirable. Vulture hock, or long, stiff feathers growing back and down from the hock are quite common and must be avoided. To prevent this, breed from birds whose hock feathers are soft and curl around the joint, and not from those whose hock feathers are quilly and stiff. It is often difficult to get good length of toe feathering without getting vulture hock.

It is generally considered desirable to feed the birds on soft feed, without much hard grain, in order to produce the soft feathering desired. This is just the opposite of the feeding used for games, where hard feathering is wanted. The birds which are intended for exhibition should also be kept rather quiet and not allowed to run or to scratch too much for feed, as this tends to wear off the toe feathering.

In breeding Cochins, the following defects must be guarded against in so far as possible: stiltiness, especially in males; not well-feathered underneath; too large combs; light eyes; too long and too stiff main tail feathers; too scanty shank and toe feathering; a V-shaped junction of the leg and toe feathering; vulture hock; too small size.

The Partridge Cochin

In breeding this variety, a single mating may be used as follows:

Select a male whose hackle, back and saddle feathers show clear, distinct black centers. The red edging should run clear around the ends of the feathers. Smut frequently occurs in hackle and must be looked out for. The wing bow wants to be a clear, bright red, free from black smut. The back should be a trifle darker than the wing bow. It is this clear red wing bow which gives the beautiful mahogany ground color sought for in the female and which cannot be obtained without such a wing bow in the male. It is also the clean black striping of hackle and saddle of the male which gives the clean penciling in the female. Any shafting in these sections will show as shafting in the females, particularly in the breast. A male having some red peppering in fluff is not objected to as a breeder, as such a male tends to strengthen the mahogany penciling in the females. Such a male usually has the finest of under color, while one with a clear, black fluff is apt to be weak, that is, be too light in under color. A little red edging in the breast is not objectionable. A good, sharp red edge on the wing flights is desirable, as a bird possessing it is more likely to get good, clear penciling in his pullets. There is a tendency for the under color to run light or white, particularly in hackle, back and saddle, which must be avoided. The black sections of wing and tail must also be free from purple barring.

The females should be a beautiful, clear, mahogany ground color, distinctly and regularly penciled with black. Often the penciling is inclined to be mossy, that is, the black and mahogany are not distinct, but tend to run together or mix to some extent. It must be remembered that females improve in penciling with age, and that pullets which are not quite clear should not, therefore, be discarded until they have molted in as hens. The best penciled pullets are never as good in this respect as the best hens. The best penciling is found in hens which are three or more years old. The strongest and best penciled females are likely to show a peppering of red in wing flights and secondaries. Pullets also never show the length of feather which hens get.

In breeding this variety, the following defects in addition to those common to the breed (page 147) must be guarded against in so far as possible: indistinct black centers or bay or red shafting in the hackle, back and saddle feathers of males; smut in hackle of male; smut in wing bows of male; white or light under color in hackle, back and saddle of male; purple barring in black sections of male; mossy or indistinct penciling in females.

For information regarding double mating for Partridge color, the reader is referred to the material on mating Partridge Plymouth Rocks (page 92).

The Buff Cochin

In mating this variety, it is unnecessary to resort to double mating, as high class standard specimens of both sexes can be procured from a single or standard mating. Double mating, as in other buff breeds, is, however, sometimes employed. In mating this variety the same considerations of color apply as in the Buff Plymouth Rock (page 89), and the matings should therefore be selected upon the same basis.

For defects to guard against, see the general description for the Cochin (page 146) and the color defects as described for the Buff Plymouth Rock (page 89).

The White Cochin

In breeding this variety only the standard or single mating is employed. Birds of both sexes as near standard as possible are selected. In color they should be pure white. The most serious color defect is brassiness. This is most apparent in males and shows as a brass or yellow color to the surface. Do not breed from brassy males

or females from a brassy flock.   Creaminess as shown in
the under color or on quills is less troublesome, but
should be guarded against.   The creamy color due to immature or sappy condition of feathers must not be confused with the creaminess apparent in mature feathers.

The Black Cochin

This variety is probably the poorest of the Cochins, as
it tends to run small and also not of as good type as
either the Buff or the Partridge.   It also tends to be
shorter and closer feathered.   In breeding, only the
standard or single mating is employed.   Select birds
which are as near standard both in type and color as
possible.   White frequently occurs in the under color
of hackle, back and saddle, especially the hackle of males,
and must be selected against.   Red, especially in the
hackle, sometimes occurs in males.   Such males should
not be used for breeding unless the color of the females
tends to run out and to become a brownish or dull black.
If this happens a male with a little red in hackle will help
to restore the lustrous black color.   Purple barring must
be avoided.   White sometimes shows in the sickles of
males.   This frequently develops in males with age, and
consequently is a more serious fault in cockerels than in
cocks which were free from it as cockerels.   Cockerels
showing this defect should never be used as breeders if
it can be avoided.   The wing flights frequently show a
little gray.   Birds without this are superior as breeders.
Care must be taken to see that no birds are used as
breeders which do not show yellow bottoms to the feet,
as birds sometimes occur in which this section is white.
As in other black breeds, the young chicks are apt to
show white in their chick feathering.   This is usually
lost as the mature feathering is gained, and if so is of
no importance.

In breeding this variety, the following defects in addition to those common to the breed (page 147) must be guarded against in so far as possible: small size; too short and too close feathering; white in under color of hackle, back and saddle, especially hackle of males; red in hackle of males; brownish or dull black surface color in females; purple barring; white in sickles of males; gray in wing flights; white bottoms to feet.

The Langshan

In type, the Langshan is quite as distinct as the Cochin, but is in many respects just the opposite. While the Cochin is very low set, the Langshan is tall and is set rather high on legs, being considerably higher than the Brahma. The body tends to be rounder than the Brahma, which is due to the fact that it is not so long and that the length and depth of body are nearer equal. It is important that the depth of body should be good and that the breeders do not show deficiency in that respect. The height of the Langshan is due to its good depth of body as well as to its long legs and the upright carriage of tail and head. However, its legs should not be so long as to make the bird appear stilty. It is lighter in weight than the Brahma and Cochin.

The back is rather short, but the high carriage of tail and head makes it appear shorter than it really is. There is a tendency for the back to be a little too long. The back, tail and neck form a U which is a trifle wider at the top than at the bottom. This U shape of back is more marked and perfect in the male than in the female. The back should be broad.

The breast should be round, full and broad and should be carried well up. The Cochiny type of breast, which is much looser in feather, very full and carried much closer to the ground, should be avoided. Narrow breasts which are associated with legs set too close together and flat breasts, especially in males, should be avoided.

The comb, which is single, should be medium in size. There is a tendency for the comb to be a little coarse, that is, too high and too large. There is also a tendency for the comb of the female to lop. Both these defects must be selected against in the breeders.

The neck is long and very erect, having somewhat less arch than either the Brahma or the Cochin. The hackle and back should blend smoothly so that no angle is formed between the two sections.

The carriage of the tail is distinct among the larger breeds and has much to do with giving the Langshan its characteristic shape. It is carried very high, but must not be so high as to be squirrel tailed. The tail should be well spread, and the junction of the tail and back must be smooth and free from any angle. Before the tail is fully grown, it may sometimes appear to be too low, but in well-bred stock, as it develops, it will usually prove to be high enough. Some pullets or hens before the full tail is grown may seem to show some tendency toward a cushion, but as the tail comes in these feathers seem to fill in at the base of the tail and do not appear as a cushion, but serve to give a smooth junction of the tail and back, free from any angle. Such females are especially valuable breeders of males, as they breed profuse saddle and back plumage, and produce males with a smooth union of back and tail free from any angle.

Since the Langshan should be well up on its legs, the legs must be long. In fact, legs of equal length in any other breed would give the bird a stilty appearance. However, the length of leg does not want to be carried to such an extreme that the birds are out of proportion. On the other hand, birds should be avoided as breeders which are too low set or too squatty, that is, have too short legs and which are a little small all around, even though they may be good in type, as they are too much on the pony order. The hock joints should be distinctly visible and should not be covered by the fluff and body feathers, as in the Cochin. There is a tendency for birds of this breed to have weak legs and hock joints. Such birds do not stand strongly on their legs, but tend to wabble and teeter. Legs or hocks set too cose together must also be avoided.

The feathering of the Langshan is quite hard and compact, more so than the Brahma. Loose-feathered birds should be avoided as breeders.

The feathering of the legs and toes is much lighter than that of the Brahma. The middle toe, unlike both the Brahma and Cochin, should be bare. Slight vulture hocks or stiff feathers in the hocks occasionaly occur, but they are not so frequent or so troublesome as in the Brahma and especially the Cochin.

The following defects must be guarded against in so far as possible in breeding Langshans: body lacking in depth; too short backs; too long backs; Cochiny breast, that is, one which is carried too low; narrow breast; flat breast, especially in males; too large and coarse comb; lopped comb in females; squirrel tail; too short legs or too low-set or squatty birds; too long legs; weak legs and hock joints; too loose feathering; middle toe feathered; vulture hocks; birds which are of good type, but a little small all around, so that they are on the pony order.

The Black Langshan

In breeding this variety, a single or standard mating is used. Birds of both sexes are selected as breeders which approach as closely to the standard as possible. Select a male free from foreign color and showing a rich, green sheen as free from purple barring as possible.

Usually the birds which are the best in color as cockerels and pullets, that is, which have the best green sheen and are free from purple barring, prove to be the best colored cocks and hens after they have molted. Occasionally a young bird free from purple will show it after the molt, while another showing some purple may molt in free from it. Purple barring, therefore, seems to be largely a matter of breeding, but a bird's condition during the time it is growing its plumage undoubtedly has something to do with it. The under color should be black, and a light or gray under color, especially in back and hackle, must be looked out for. It is better to have a surface color showing a green sheen with some purple and good under color than a dull, dead black surface color with light under color. The color of the female should be the same as the male, except that it will not be so brilliant in green sheen. The wing feathers sometimes show frosted tips, that is, white or gray tips. This is a defect which must be guarded against. There is some tendency in males toward a bronze tinge on the shoulders and especially on the tail. While this defect must be selected against, it is not so serious as purple barring. Sometimes males come with red or straw color in hackle, back or saddle. Some breeders use these males to improve the green sheen in the offspring. Other breeders do not consider it necessary to use such males to secure good green sheen, and cull all males showing this foreign color. As in all other black varieties, the chick plumage is apt to show some white feathers, which usually disappear, however, when the adult plumage is gained.

A black or dark brown eye is desired, but there is a considerable tendency for the eye to be too light, that is, red or yellow. This tendency must be carefully selected against.

There is also a tendency for the bottoms of the feet to

be yellow. As this disqualifies, it is important to use birds the bottoms of whose feet are pinkish white.

Briefly summarized, the defects in addition to those common to the breed (page 153) which must be guarded against in so far as possible in mating this variety are: too light eyes; gray or light under color of back and hackle of males; yellow bottoms to the feet; purple barring; dull, black surface color lacking the green sheen; frosted tips to the wing feathers; bronze tinge on the shoulders and especially on the tails of males.

The White Langshan

This variety is not, on the average, quite as good in type as the Black variety. In breeding, it is universal to employ the single or standard mating, selecting breeders of both sexes which are as near standard, both in type and color, as possible.

As in other white varieties, the birds should be used as breeders which have the purest white plumage. They should be free from creaminess in both sexes, and from brassiness in males. The plumage both in quill and web should be free from black ticking and from any black feathers in any section. The legs should be slaty in color and a tendency for them to run too light must be avoided.

Briefly summarized, the following defects in addition to those common to the breed (page 153) must be guarded against in so far as possible in mating this variety: brassiness; black ticking; black feathers; too light color of legs; creaminess.

CHAPTER VI

THE MEDITERRANEAN CLASS

The Leghorn

The type of all the different varieties of Leghorns should be identical. The birds should be characterized by a smoothness, a style, and an alertness, all of which combine to form a breed of rare beauty and attractiveness. There should be no suggestion of heaviness or sluggishness. The birds should be free from any angular appearance, the body shape being f o r m e d of s m o o t h, sweeping c u r v e s throughout. Birds of both sexes should have a rather high station, showing a good length of shank, while the hock joint and a part of the thigh should be distinctly visible. See Fig. 55. There is a tendency, particularly in males, for the birds to be too low-bodied or too short-legged. This is sometimes due to the settling which may occur with age. When this condition

Fig. 55—Single Comb White Leghorn cockerel of excellent type. Notice the good length of shank and that the hock joint and a part of the thigh are distinctly visible. (Photograph from the Bureau of Animal Industry, United States Department of Agriculture.)

is due to age, allowance must be made for it in selecting breeders, and a knowledge of the type of a bird as a cockerel is valuable in this connection. See Fig. 56. There is also a tendency on the part of some birds to be too upstanding and

Fig. 56—Single Comb White Leghorn cock. This is the same bird at 5 years of age as shown in Fig. 55 at one year of age. Notice how he has settled with age. The difference in the appearance of the comb is due to the points and blade having been slightly frosted. (Photograph from the Bureau of Animal Industry, United States Department of Agriculture.)

to approach the game type. Such birds invariably lack breast and are likely to carry pinched tails. The back in both sexes should be nearly level, but there is a tendency, especially in males, for the back to show too decided a slope downward from the shoulders to the tail, and for the wing points to be carried too low, so that the wing does not assume a nearly horizontal position, as it should. See Fig. 57.

Leghorns should have well-spread, medium, low-carried tails, and there should be no angle where the back and tail join, but these sections should blend in a long, sweeping curve. High-carried tails are undesirable and of necessity

form an angle between the tail and back. See Fig. 58. Where tails are low-carried there is a tendency for the birds to have a drawn-out appearance at the base of the tail, or, in other words, to lack depth at the point where the tail and body join. A well-spread tail in both sexes, and abundant long saddle feathers in males, eliminate this appearance. A well-spread tail usually denotes good health and good breeding condition. See Fig. 59.

Fig. 57—Single Comb White Leghorn male showing too decided a downward slope of back from shoulders to tail. Notice also that the wing point is carried low and that the breast is too prominent and carried too high so as to approach the game or pouter pigeon type. (Photograph from the Bureau of Animal Industry, United States Department of Agriculture.)

The well-spread tail of the males should be fully furnished. The sickles should be long and carried well up over the main tail feathers. Sickles which are too long hang down too far and have a bad appearance. The saddle feathers should be abundant and long, and coupled with a broad back and shoulders.

A full, round breast is desired in both sexes, and is a section that is likely to be deficient, both in males and females.

The head of the Leghorn is very important, and particular attention must be given to the comb and head points. The Leghorn comb is of two kinds, single and rose. The single comb of the male should be neat as a whole, that is, neither too large nor too thick. It should be of fairly fine texture, but should have a good base so that it sets well and firmly on the head. The

Fig. 58—Single Comb White Leghorn female showing high-carried tail. (Photograph from the Bureau of Animal Industry, United States Department of Agriculture.)

comb should be erect and should in no way have the appearance that it may turn over or lop. See Fig. 4, head 2. It should have five points and care should be exercised to select, if possible, breeders having no greater number of points. Some birds will, however, carry a six-point comb well, as it will be in proportion to the bird. However, it is better to select a bird with four points rather than six, if such a choice is necessary, as there is a greater tendency for the comb to carry more than five points rather than fewer. A comb the points of which are blunt or rounded, and which are of about the same width throughout their length, often looks better with six points, while one in which the serrations are pointed and taper throughout is more symmetrical with five points. The serrations, or spaces between the points, should be cut down to an even

l e v e l, see Fig. 60, and the serrations should be even or set p r oportionately the same distance apart, that is, the greater distances should be between the higher p o i n t s and the smaller ones between the shorter points, so that the entire comb will have a symmetrical a n d balanced appearance. The blade should be carried straight off the back of

Fig. 59—Single Comb White Leghorn female showing pinched tail. (Photograph from the Bureau of Animal Industry, United States Department of Agriculture.)

the head and should show no tendency to follow the neck. See Fig. 2, head 6. Most blades show some notching in the rear, but one free from this is desirable. See Fig. 60. The comb should also be free from thumb marks and wrinkles along the base. See Fig. 4, head 1. It should not be hollowed out along the side, just above the base, but should be smooth and firm.

The shape and general character of the single comb of the female should be the same as that of the male, except that it is lopped. Particular attention should be given to the

fact that the front of the comb and its first point should
stand erect, while the remainder lops to one side. The
Leghorn female's single comb differs from that of the
Minorca, which does not stand erect, but folds back over
the bill, the whole
comb lying flat on the
head and side of the
face and following
the back of the neck.
See Fig. 3, heads 4
and 5. The Leghorn
female's single comb
should be fine in text-
ure, never coarse. A
comb which is rather
undersized is prefer-
able to one which is
large, c o a r s e and
beefy. The comb of
the female tends, as
the hen grows older,
to become smaller and
stand erect. This
must be given due
consideration in se-
lecting breeders. The

Fig. 60—Head of Single Comb White Leghorn
male showing notched blade and serrations cut
down to an uneven level. (Photograph from
the Bureau of Animal Industry, United States
Department of Agriculture.)

male comb of the rose comb varieties should be of medium
size and be securely and evenly placed on the head. It should
be square in front and the sides should be carried back nearly horizontal before they begin to taper to the spike, giving
an outline to the comb something like that of a hairbrush. The
spike should be well developed and should extend straight
out from the head, with no tendency to follow the neck like
the Wyandotte comb, see Fig. 2, head 4, or to incline upward
like the Hamburg. Fig. 2, head 8. The center should be
filled out and free from hollows. See Fig. 4, head 3.

The female rose comb should be identical with that of the male, except that it is smaller in proportion to the bird.

The ear lobe should be of medium size and oval in shape. It should never be so large as to be at all pendulous, nor should it be large and round, like the lobe of the Hamburg. A large lobe is more likely to be accompanied by white in face than a smaller, neater lobe. It should be as smooth and free from wrinkles as possible. It should also be white. Often the lobe, especially in males, shows considerable red peppering, which is more commonly found in older birds, as it tends to develop with age. This peppering must not be considered as serious a defect in older birds as in the young ones. Usually birds showing a red peppering in the lobe are free from white in face. Many fine young birds have a yellow lobe, which may vary all the way from a slight tinge of cream to a decided yellow. It is most noticeable in birds particularly strong in yellow leg color, and just before they get their adult plumage or before they begin laying. It occurs more commonly in males than in females, but fades out more quickly in females. Yellow lobes are more apt to show when the birds run in damp places, or in clover or alfalfa. It need not be considered a serious breeding defect. Red spots and wrinkles in the lobes of males are often due to injury received while fighting, or from being picked by females. Such defects, due to injury, need be given no weight in breeding.

The face of both sexes should be red. One of the most serious and troublesome defects in Leghorns is white in face, as it not only disqualifies in cockerels and pullets, but is a characteristic which is quite strongly transmitted, and once introduced into a flock, is difficult to breed out. The white may show as white extending off from the ear lobe, or as detached spots on the face, especially under the eye. The paleness of face due to lack of condition, or the white cast due to the hair on the face, especially in females, must not be confused with white in face. Cocks and hens with age often develop some white in face, although they were sound in this respect as young birds. If the amount shown is small, this is not considered as serious a defect for breeders, provided their faces were sound as young birds. However, cocks and hens which remain sound in face with age are preferable as breeders. In a good sound face the line between the white of the ear lobe and the red of the face should be clean-cut, sharp and well defined.

As a breed, the Leghorn is very free from side sprigs, stubs or down.

Common defects of the Leghorn, which should be guarded against in breeding, in so far as possible, are: too many points to the comb (in single combs); thumb marks on comb (in single combs); double points on comb (in single combs); for large, beefy comb (apt to lop in males); too heavy blade to comb (apt to turn to one side); white in face; lack of breast; tails carried too high; wry tail, that is, not carried straight, but turned to one side; pinched tail, that is, tail not well spread, especially in females; pearl or light eye; notched blade to comb; back showing too great a slant downward from the shoulders to the tail; wings carried too low.

The Single Comb Brown Leghorn

In this variety, in order to secure standard specimens of both sexes, it has been necessary to resort to two separate matings which are entirely different in character. Breeding along these two different lines has been carried on to such an extent that two entirely different sets of blood lines have been established, which, in the opinion of many breeders, are so distinct as to be virtually distinct varieties. In fact, there is a movement on foot at the present time to bring about a division of the Brown Leghorn variety, as it is now included in the standard, into two separate varieties, one of them, which corresponds to the cockerel-bred line, to be

known as the Dark Brown Leghorn, and the other, which
corresponds to the pullet-bred line, to be known as the Light
Brown Leghorn.*

Fig. 61—Well-marked Brown Leghorn feathers from a Dark or Cockerel
bred male, i. e., an exhibition male and from a light or Pullet bred female,
i. e., an exhibition female. M indicates male and F female. (Photograph from
the Bureau of Animal Industry, United States Department of Agriculture.)

The Dark Brown Leghorn, or cockerel mating.—Select a male which is as near standard as possible. Care must be taken to see that the red of the neck, wing bows, back and saddle is rather deep and very rich in tone, and that the shade of color of these sections matches, or is as nearly the same as possible. Often the red of the hackle runs into a lighter shade at the ends of the feathers, giving a light cape and causing what is known as a shawl effect. This is undesirable. Select for good, sharp striping of sound, metallic black in hackle and saddle, and for a border of red which extends clear around the end of the feather. Cockerels showing slightly pinched tails often molt in with much better spread of tail as cocks, and in consequence, this defect must not be allowed to count too severely against an otherwise good cockerel. All black points should show a greenish sheen.

Select females for this mating which are bred from a line of Dark or exhibition males. In type these females should be standard, but may be slightly more masculine in appearance. The combs should be firm, low set and preferably erect, the serrations firm and even, and the combs should not carry more than five points. Combs carrying four points are preferable to those with six, as the tendency is for the males to come with more rather than less than five points. Good red eyes are desired and faces free from any trace of white. The lobes should be sound and white, yellow lobes being carefully avoided. Females from a line of Dark or exhibition males will have coarse stippling in back and wings and will be darker in color than the Light or exhibition females, showing a greenish cast over the back and tail. The hackle feathers should show a clean striping of greenish black, which should be bordered entirely around

* "At the Annual Meeting of the American Poultry Association at Chicago in August, 1919, the Light and Dark Brown Leghorns were admitted to the Standard as two separate varieties in place of the Brown Leghorn as a single variety."

the feathers with a reddish orange. This will result in a hackle having a decidedly reddish effect, and any tendency toward a lighter or yellow shade should be avoided. Females with breasts stippled or penciled similar to the back are used and give good results in breeding. The wing flights and secondaries should be just short of positive black, the two outer primaries of each wing showing a slight reddish tinge. The under color of all sections should be slate, or even dark slate, clear to the skin. The shanks should be as yellow as possible, but a slight tendency to dark brown is not a serious objection.

The Light Brown Leghorn, or pullet mating.—The females used in this mating should be as near standard as possible. Selection should be made to secure as perfect a match as possible in the color of back and wing bows, and to have both sections as free as possible from shafting and red or brickish color, as these are the greatest difficulties in breeding females of this variety. The striping of the hackles is apt to be weak, showing some penciling, particularly in the best stippled females. This should be given attention in selecting the females for this mating. The breast should be a clear salmon, absolutely free from any penciling or stippling in the center.

The male to use in this mating should be out of a Light or exhibition female of fine quality. Such a male will be much lighter in color than the Dark or exhibition male. The male's hackle feathers should be bordered by a light yellow or straw color, which should run clear around the end of the feathers. The stripe should be as clear a black as can be secured. The wing bows should be a light orange or light red. The saddle should approach as closely as possible a light yellow or straw color. The saddle should also be devoid of striping, but a tendency toward a slight stippling in place of a stripe is an advantage. The tail should be black, free from purple, but there is no material advantage in a greenish sheen to the tail. The breast and body should be dead black, the fluff grayish slate. A mixture of reddish-tipped feathers in breast, body and fluff is, however, desired. The under color of all sections should be slate. The shanks should be a bright, clear yellow.

The comb of this male should be similar in shape to that of the Dark or exhibition male, but may be a trifle larger. A face and wattles slightly effeminate in character is advantageous. The eye should be the same as that of the Dark or exhibition male.

In breeding this variety, the following more or less common defects, in addition to those common to the breed (page 163), must be guarded against, in so far as possible. In the Dark Brown Leghorn, or cockerel-bred male, which is the standard male: undersize; too large comb or wattles; the hackles running into lighter shade or too dark shade at the ends and producing a shawl effect. Such hackles result in what is known as a three-colored bird, or one having a different shade of color in hackle, saddle and wing bows.

In the Dark Brown Leghorn, or cockerel-bred females: too coarse combs; combs with more than five points; yellow lobes; yellow shade in hackle; too light under color.

In the Light Brown Leghorn, or pullet-bred male: pearl or light-colored eyes; extreme coarseness in comb and other head points; wing bows too dark in color; short backs; black in saddle and back; pinched or gamy tails; birds having fullness of breast, but with the back line extending downward, as such males breed too long-backed females.

In the Light Brown Leghorn, or pullet-bred females, which are the standard females: unevenness in color of back and wings; shafting in back and wings; penciling in the striping of hackle; penciling or stippling in the center of breast; brickiness on wings.

The Rose Comb Brown Leghorn

In mating the Rose Comb Brown Leghorn, the same general principles must be observed, with respect to color and type, as in the Single Comb variety (page 163). In general, the striping is not so good as in the Single Comb. The line between the black stripe and the red edging is not so sharp, with the result that the striping is not so distinct. Breeders as good in this respect as possible should be selected to keep up the progress which has been and is being made in attaining an excellence of striping equal to that of the Single Comb variety.

In comb, it is necessary to guard against those which are too broad or too high and those in which the spike shows an inclination to follow the neck. Combs with hollows in the sides or in the top or center must also be avoided.

The Single Comb White Leghorn

In mating the Single Comb White Leghorn, it is usual to employ standard matings. Both the male and the females are selected to conform as nearly as possible to the standard description, as given in the American Standard of Perfection. Where weakness is shown in any section or sections of either sex, it should be offset by strength in those sections in the opposite sex. Attention and care must be given to the general considerations for the breed which have been described.

In selecting breeders for color, use those which are the whitest. The White Leghorn is quite free from any black ticking or other foreign color. Brassiness sometimes occurs in males, and, if present, is to be found on the shoulders, back and saddle, showing as a yellow or brassy cast to the surface color. Brassy birds should never be used as breeders. Creaminess is more common than brassiness, but is not as serious a defect. It shows as a slight yellowish cast in the under color and in the quills of the feathers close to the body. Creaminess is especially prevalent in young stock, while their adult feathers are still green or sappy. Occasional males show a slight tinge of red, usually on the shoulders. Occasional females show a slight tinge of buff or salmon on breast, on head, or on both. Such individuals should be discarded for breeding unless they are unusually outstanding in other respects.

The yellow shank color of practically all White Leghorns is lost to a large extent with age. This is particularly true of females which lose the color with heavy laying. Their yellow color is renewed somewhat, though seldom completely, at each molt.

Many males also have pale legs, losing the yellow color with age. Bare yards and some soils also seem to have a bleaching effect on yellow legs. The lack of yellow in the shanks of either males or females need give the breeder small concern, unless they lacked this color as cockerels and as pullets. Males having very pale yellow legs, some of them being practically white, usually have the whitest plumage, especially free from any trace of creaminess. Such a male, especially if it comes from good yellow-legged stock, may be particularly valuable to use on a flock which shows a tendency toward creaminess.

Sometimes a mating is made which is not strictly standard, but is really a double mating within a single pen. This is done by selecting two classes or types of females to breed to the same male with the expectation of producing males of exhibition quality from one class and females, or both males and females, from the other. In making this mating, a male is used as near standard as possible. The first class of females selected to go with this male should be those which as pullets had combs which tended to stand erect and which are coarser in texture than that of the exhibition female. This type of comb is selected with the expectation of securing better combs on the males produced. These females should also be as strong in prominence of breast as possible, in order to secure males which are not deficient in that respect. They should also be selected for particular fullness across the base of the tail, showing, if possible, some cushion, and for breadth of back in order to secure good saddles and tail furnishing in the males produced. Females showing an unusual abundance of tail coverts, which extend well up on the main tail feathers, are especially valuable in producing males with fine, well-furnished tail and finely finished saddle. This part of the mating is expected to produce cockerels of good quality, while the pullets are not so likely to be of good quality.

The second class of females selected for this mating should be as near standard as possible, and these are depended upon to produce pullets of good quality, and perhaps some cockerels also.

The common defects of the variety, in addition to those common to the breed (page 163), to guard against, in so far as possible, are: brassiness; creaminess; red or buff or salmon in the plumage; yellow or too coarse ear lobes; too coarse and too large body; the game and pouter pigeon type, that is, the breast too prominent and carried too high. See Fig. 57.

The Rose Comb White Leghorn

The mating of the Rose Comb White Leghorn is identical with that of the Single Comb variety, except in the matter of comb. Combs which are too broad or too high, those in which the spike tends to follow the neck, and those showing hollows along the sides, or in the center or top, must be avoided in so far as possible.

The Single Comb Buff Leghorn

This variety tends to have especially large, coarse, beefy combs which may be badly thumb-marked. Neat-headed breeders are therefore valuable. The ear lobes and faces tend to run quite good and do not give much trouble.

In mating this variety, as with other buff breeds, it is not necessary to resort to double mating in order to produce high class standard specimens of both sexes, but double mating is sometimes employed. In the matings the same color considerations will guide in the selection of the birds for breeders as in the Buff Plymouth Rock (page 89). However, it should be kept in mind that the Buff Leghorn is not troubled with mealiness to the same extent as the Buff Plymouth Rock. In case a double mating system is used, the male for the pullet mating should have a rather small comb and his head may be less masculine in character than the male in the cockerel mating, in order to offset the tendency toward large combs on the females.

In addition to the common defects for the Leghorn in general (page 163), the following defects must be especially avoided in this variety, in so far as possible: large, coarse, beefy combs in both sexes; thumb-marked combs in both sexes; comb of females not erect in front, but folded and flat on the head like the comb of the Minorca female; too high carried and squirrel tails; pinched tails, which are especially troublesome in females; poorly furnished tails in males.

For color defects to be guarded against, see those of the Buff Plymouth Rock (page 91).

The Rose Comb Buff Leghorn

In regard to color, type and general considerations, the mating of Rose Comb Buff Leghorns is identical with that of the Single Comb variety (page 170), except for the comb. Combs which are too broad or too high, those with the spike following the neck, and those in which the spike is absent, must be guarded against in both sexes. Combs showing hollows in the sides or hollows in the top or center must also be avoided.

The Black Leghorn

In mating the Black Leghorn, a single or standard mating is almost universally used. Both the male and females should be as near standard as possible. Birds should be chosen which have a green sheen to the black and are free from purple. The under color should be black throughout, and care must be exercised that the male does not show white in the under color of hackle, back and saddle. It is especially hard to eliminate white from the under color of hackle. White in under color is a defect which tends to grow worse with age, and consequently cockerels sound here are greatly to be preferred to those which show the defect even slightly. Old males sound in these sections are especially valuable breeders of black under color. White is sometimes found in the sickles of cocks. There is not much trouble from white in the wings, although the young stock often show white in chick wing feathers and also purple until they get their adult plumage. While dusky yellow legs or shanks are allowed, yellow is preferred, and good yellow legs can be secured in this breed without sacrificing black under color. The birds of both sexes chosen for breeders should be strong in color of face and ear lobes.

Defects which are common to this variety, in addition to those common to the breed (page 163), and which must therefore be guarded against, in so far as possible, are: too high and too coarse combs; too many points to combs; thumb-marked combs; tails carried too high; white in face; white in sickles of cocks; white in under color of hackle, back and saddle of males, especially hackle; purple sheen in females; black legs; red in plumage, especially males.

The Silver Leghorn

In mating the Silver Leghorns a single mating is used. The male should have a silvery white saddle and the hackle should show a faint stripe only in its lower portion. It is important that the breast be black, without any white appearing in it. A slate under color is desired. The best males for breeders show a little frostiness in the fluff. The best breeding males show in their chick feathers some stippled feathers under the wings and in the tail coverts. The females used as breeders should be as near standard as possible. Penciled females must be avoided, as they produce males with a stripe in hackle and saddle.

The following defects, in addition to those common to the breed (page 163), must be looked for and guarded against, in so far as possible, in breeding this variety: too large and too coarse combs; thumb-marked combs; red in ear lobes; breast feathers of males tipped or edged with white; a tendency to brassiness in saddle and back of males; brick or reddish color, both sexes; shafting in back, shoulders and wings of females; red in wing bows and shoulders of males.

The Red Pyle Leghorn

It is common practice to double mate in this variety, as more satisfactory results are thus obtained.

Cockerel mating.—Use a male for this mating which is as near standard as possible, and give special consideration to securing good, strong red coloring of the secondary wing feathers. The females should have a solid salmon breast and should also show a tendency to reddish color on the lower part of the wing bows.

Pullet mating.—The male selected for this mating should be slightly lighter in color than the exhibition male. It is also rather an advantage if the hackle is slightly striped with red. The females used in this mating should be as near standard as possible.

In addition to the common defects which should be guarded against for the breed in general (page 163), the following defects must be especially avoided, in so far as possible, in mating this variety: too large and too coarse combs in both sexes; thumb-marked combs in both sexes; blade of comb following the neck in males; comb of females not erect in front, but folded and flat on the head like the comb of the Minorca female; red in ear lobes; too high carriage of tail; salmon on wings of females; black in plumage, especially males; legs too light in color, approaching white; breasts of females too light in color.

The Minorca

The Minorca is characterized first of all by its size. It is the largest breed of the Mediterranean class. The Single Comb Black Minorca is the largest variety of this breed, averaging about a pound heavier than the other varieties, which are of about the same size as the White-Faced Black Spanish and are larger than any of the other Mediterranean breeds. The type of the Minorca is quite distinct. It is an upstanding bird with a long body which has a gradual slope downward from front to rear. This is especially noticeable in the back line, which is long and straight but slopes downward from the shoulders to the tail. The tail is of good size, well spread, and low carried. The Minorca has not the smoothness which the Leghorn displays and has in general a more angular appearance. There is sometimes a tendency shown for the back in both sexes to be curved or arched instead of straight, but this is not very troublesome. A round, full breast is most desirable in both sexes. The birds should stand firm and straight on their legs. There is a tendency toward knock-knees in this breed. When birds which are too high and too long-legged are mated, this knock-kneed condition is one of the most troublesome defects with which one has to contend. The wattles are long and the large white ear lobe desired is an important point of the head, as it adds much to the appearance of the breed. Red in ear lobes is a troublesome defect which must be guarded against in all varieties of Minorcas. White in face may occur, especially in males, and tends to increase with age. See Fig. 62. Young birds showing white in face should not be used, and if old males can be secured which are sound in this respect, they are to be preferred. However, a cock showing white in face may be used if he was sound in face as a cockerel. It is important to select good-sized, rugged birds to avoid weak constitutions. Good size and bone can be secured by m a t i n g a large male to average-s i z e d females, or vice v e r s a. S l o w maturing birds should never be used for breeders. Stubs occasionally occur in Minorcas a n d must be selected against.

The comb of the single comb varieties is large a n d r a t h e r coarse in texture and deeply serrated. It has six points instead of five, as in most other s i n g l e c o m b fowls. See Fig. 2, head 5. The comb of the male should be erect and with a strong, thick base to overcome the tendency to be weak and lop or turn over on account of its large size. Females which have a good thick comb base are more apt to get good combs in cockerels. The blade of the male's

Fig. 62—Single Comb Black Minorca Cock showing white in face. (Photograph from the Bureau of Animal Industry, United States Department of Agriculture.)

comb tends to follow the curve of the neck. See Fig. 2,
head 5. It is important that the comb points in both sexes
be wedge-shaped.

The single comb of the Minorca female is large, six-
pointed, and in shape differs distinctly from the Leghorn
female comb. It does not stand erect in front as in the
Leghorn, but the front lops to one side across the beak, then
loops or folds and falls with the remainder of the comb to
the other side of the head. The whole comb lies flat on the
head and hangs down over the face, no part being erect.
See Fig. 3, heads 4 and 5. Breeders differ in the type of
comb used in the females, some using females with larger
combs and some with finer combs. In general, it may be
said that the females with the larger combs will produce the
best combed males, and those with the finer combs the best
combed females.

The rose comb in both sexes is quite large, with a prominent spike which follows the curve of the neck to some
extent. There is a tendency for the rose combs to be too
high and too broad and to show hollows, both along the
sides and on the top in the center.

The following are defects which are characteristic of the
breed and must be guarded against in so far as possible:
knock-knees; too high tails; pinched tails; weak constitution; white in face, especially males; curved or arched
backs in females; too short backs; stubs; too high, weak
single combs in males, which are apt to turn over; lopped
single combs in males; too high or too broad rose combs in
both sexes; hollow rose combs in both sexes; red in lobe,
unless due to injury or exposure to the weather; slow
maturing birds.

The Single Comb Black Minorca

In mating this variety it is customary to use a single or
standard mating, although females varying in certain

respects are often used in the same mating, some primarily for the production of cockerels and others primarily for the production of pullets. A male should be selected which is as near standard as possible, being careful to see that the breast is round and full and that he is strong on his legs and not knock-kneed. Also see that his comb is strong and shows no tendency to lop. Males showing a few red feathers in the hackle often make good breeders, but any which show a tinge of red on the wing bar should be discarded. He should be as free as possible from purple barring. A good, large, dark eye in the male is especially important. Such a male will get a majority of dark-eyed cockerels and pullets, but a red-eyed male, even when mated to dark-eyed females, will usually produce a majority of red-eyed cockerels and pullets.

In selecting the females, be sure that all have round, full breasts and that the back is not curved or arched. The tails should all be well spread, the wider the main tail feathers the better, but the carriage of tail may vary considerably, some being low and others considerably higher. The carriage of the tail of the females will also depend on the tail carriage of the male. If his tail is carried low, some rather high-tailed females can be used, but if his tail carriage tends to be high, low-tailed females should be used.

In color, it is best to secure dull-colored females if possible. However, if the females show some purple which comes from neglect or exposure to the weather, they may be used. Some females showing purple as pullets will molt in good as hens, with proper care. Some females which had good-colored flights as pullets are apt to show some white as hens. This is not serious, however, and such hens can be used in the breeding pens if they were sound as pullets.

Females showing some variation in length of shank may be placed in the breeding pen, some being longer or shorter in this respect than others. Some variation of comb is also desirable. Some of the females should have large combs and good thick comb bases, as these will produce the best combs on the cockerels. Others should have finer combs, as they will produce the best combs on the pullets.

In addition to defects common to the breed (page 176), the following must be guarded against in so far as possible: purple barring; red eyes; red tinge on wing bows of males; white in flights of females.

The Rose Comb Black Minorca

This variety is identical with the Single Comb Black Minorca except in the matter of comb, which is rose, and in size. The same points already covered for the Single Comb variety (page 176) and for the breed in general (page 174) must be taken into consideration in breeding this variety.

The Single Comb White Minorca

Single or standard matings are used in this variety, both sexes being selected as near standard as possible. While this variety should have the same type as the Black Minorca, in size, like the Single Comb Buff and the Rose Comb Black varieties, it runs slightly smaller than the Single Comb Black. Even with this difference in size, however, there is a tendency for the birds of this variety to be too small and it is therefore necessary to select individuals which are of good size. The plumage should be pure white throughout and brassiness must be avoided. Select for pinkish white shanks, as there is a tendency for the shanks to have a bluish cast.

In mating, guard against the following defects, in so far as possible, in addition to the defects common to the breed (page 176): blue cast on legs; brassiness; too small size; too short backs; too high tails.

The Rose Comb White Minorca

In mating this variety the same points must be taken into consideration as in the Single Comb Whites, except in the matter of comb. In comb the same conditions exist as in the Rose Comb Blacks (page 176).

The Single Comb Buff Minorca

In mating this variety it is unnecessary to resort to double matings, as high class standard specimens of both sexes can be produced from a single or standard mating. Double mating, as in other buff breeds, is, however, sometimes employed. In mating this variety the same considerations of color apply as in the Buff Plymouth Rocks (page 89), and matings should be selected upon the same basis.

As a variety, the Buff Minorca is not as large as the Black, having the same standard weights as the white. This variety tends to be lower set on legs than the Black variety.

It is difficult to get good under color in the females of this variety, as it tends to be too light. The under color of the males is good.

This variety is very free from white in face and the combs are more medium in size than in either the Black or White varieties.

The following defects must be guarded against, in addition to those given under the general description for the Minorca (page 174), and the color defects, as described for the Buff Plymouth Rock (page 91): light eyes; yellow legs and blue legs.

The White-Faced Black Spanish

This breed is unique in appearance because of the extremely large white faces. In type these birds are much like the Minorca, but the tails are carried somewhat higher in both sexes, which gives a sharper angle where the back and tail join. The back line should be long and show a downward slant from the shoulders to the tail. In size they are only slightly smaller than the Single Comb Black Minorca.

The comb is quite large, but has five points instead of six as in the Minorca. The comb of the female is erect in front, like the Leghorn's, instead of having the double fold and lying flat on the head as in the Minorca female.

In mating this breed it is usual to employ only a single or standard mating. The very best colored birds that are strong in greenish sheen should be selected in both sexes. There is not much trouble with purple barring in the adult plumage, although it often shows in the chick feathers. Some white in flights is also apt to appear in the chick feathers and sometimes in adult plumage. Birds showing this defect in the adult plumage should not be used as breeders.

Select a male which has a good stiff comb, as there is a tendency for the comb to lop at the back. Avoid a female comb which approaches the type of the Minorca female comb. The two top feathers or sickles of females should bend or curve a trifle.

One of the most important points in this breed is the face. Use the largest and the smoothest faces possible for breeders. There is more or less trouble to get the faces large and at the same time smooth. There is a tendency for the face to become more wrinkled, rough and puffy with age, and at two years old they may cover the eyes so as to obstruct the sight. This is particularly true of males, as the faces of females never get as rough as those of males. The face generally increases in length until the birds are two years old. Some male birds are known to have good white faces measuring nine inches in length. Sometimes pulling the faces is practiced in an effort to increase their size. This is neither necessary nor desirable, as pulling causes them to become wrinkled, red and rough. Birds should also be chosen with faces as free from red as possible.

As young birds, these fowls are not particularly hardy. They must therefore have good care until they are eight or 10 weeks old, and must be kept out of the cold and damp. Spanish males, particularly the older ones, should be fed in receptacles raised off the ground 12 or 15 inches. If this is not done, they are often unable to see grain placed on the ground owing to the fact that the face may become wrinkled and nearly cover the eyes.

The following defects must be guarded against in this breed in so far as possible: too high tails in both sexes; wry tails; red in face; red in ear lobe; wrinkled, rough or puffy faces, especially in males; combs lopped at the rear in males; too large or too coarse combs; combs of females not erect in front, but folded and flat on the head as in the Minorca female comb; white in flights; white tips to main tail feathers.

The Blue Andalusian

This breed, by virtue of its blue color, is not only an unusual looking fowl, but one of decided beauty as well. The lacing of darker blue which occurs throughout the female and on the breast and body of the male adds to its attractiveness of appearance.

In type, this breed is intermediate between the Minorca and Leghorn. It is a bird of smoothness and quality, with medium low-carried tail, and standing well up on its legs. The shoulders are high and prominent, and the back line, both in male and female, is not level as in the Leghorn, but shows a downward slant from the shoulder to the tail as in the Minorca. The juncture of the back and tail are not so smooth as in the Leghorn, but should not show a sharp angle. The Andalusian has a long body, nearly if not quite as long as the Minorca. In the female the body has a decided wedge or pear shape, the deeper part being at the rear. The breast is carried high, considerably higher in this respect than the Leghorn. The back line should be straight and should show no tendency to arch, or as it is sometimes termed, to show a roached back. The comb has five points and is much like that of the Leghorn in shape, except that there is more of a tendency for the blade to follow the neck in the male. In size the comb tends to be somewhat larger than that of the Leghorn. This breed lays a white egg, but occasionally tinted eggs are secured. The presence of these

Fig. 63—Well-marked Blue Andalusian feathers. M indicates male and F female. (Photograph from the Bureau of Animal Industry, United States Department of Agriculture.)

tinted eggs is not to be taken as any evidence of impurity of the stock.

In breeding, the Andalusian manifests a peculiar and interesting occurrence. When blue birds of both sexes are mated together they do not give offspring all of which are mated together they do not give offspring all of which are blue, but approximately 50 per cent blue, while approximately 50 per cent come either black, white, or white splashed with blue or gray. Of this 50 per cent of offspring which come

Fig. 64—Black Andalusian male. (From the Kansas State Experiment Station.)

other than blue, about one-half, or 25 per cent of the total offspring, come black, while the other half, or 25 per cent of the total offspring, come white or white splashed. Red also occasionally shows in these off-colored birds. The long-continued selection of pure blue birds for breeding seems to have resulted in little or no advancement in the percentage of blue offspring obtained. If, however, a black bird of either sex is mated with a white or splashed white bird of the opposite sex, nearly 100 per cent of the offspring will be blue. With the peculiar behavior of the Andalusian in breeding there are a number of methods of mating which can be and are used.

The Blue Andalusian Club of America advocates some one or more of the following matings: (1) Standard exhibition birds of both sexes; (2) a very dark blue, heavy laced male of good type, carriage and head points to females of good type and of a very light shade of blue, even to the extent of possessing little or no lacing; (3) an extremely light blue bird of good type, carriage and head points with the very darkest, most heavily laced blue females; (4) a pure black male bird of good type, carriage and head points with splashed white females of good type; (5) a splashed white male of good type, carriage and head points with black females of good type.

Fig. 65—Black Andalusian Female. (From the Kansas State Experiment Station.)

The method of mating as given by different individuals varies somewhat, and for this reason the description of two such matings is given below. One well-known breeder gives his method as follows: Do not double mate. Select a male with good solid black or blue-black back. He should have good, distinct lacing, even the main tail and wing flights showing signs of lacing if this can be found, but males showing these signs of lacing in tail and wing are very hard to get. If a clear, silvery blue male is used, the offspring are likely to come brassy or reddish in hackle. The male in this mating is depended upon mainly for the color of the offspring. Select females of good type, which are well up on legs, with low carriage of tail and a good wedge-shaped body. These should show as clear a blue in plumage as possible, with as good lacing as can be obtained. Avoid hens showing a smut or frost of dark color throughout the body and wing.

Another breeder describes his method of mating as follows: Use a standard mating for color. That is, do not select extremes, such as the use of a very dark male with light females, or vice versa. The use of these extremes in mating is apt to cause a mottling of darker color, in the females particularly. The lacing of the birds se-

Fig. 66—Splashed White Andalusian male. (From the Kansas State Experiment Station.)

lected should be as good and distinct as possible. Where extremely dark birds are used, there is a tendency to throw red in the hackle of the male offspring. If too light birds are used, the hackle is likely to come too light or faded out in color. This breeder suggests that it is bad policy to breed a blue male which came from a black and splashed white mating, as the use of such a bird is likely to result in the loss of good lacing, and may cause a light smutty blue. Always use a blue male from a blue mating. While it is bad policy to use a blue male from other than a blue mating, it is well to reserve the black and splashed white pullets or cockerels of good type for use in the black by splashed white mating.

While this breeder does not double mate for color, he occasionally makes special matings for head points and type, such as the use of an effeminate-headed male to breed good heads on pullets.

Although a majority of breeders of the Blue Andalusian feel that it is not necessary to resort to double mating, yet

this is sometimes done with the idea that it is easier in this way to secure males with a good blue-black top color, and at the same time clear, distinct lacing on breast and throat, and to secure females which are an even shade of blue and distinct, contrasting lacing without a tendency to weaken in color.

Fig. 67—Spashed White Andalusian female. (From the Kansas State Experiment Station.)

One method of double mating this breed is as follows: Cockerel mating—Select a standard colored male which shows a very clear and distinct lacing from the throat down the entire length of the breast. The lacing must be well defined, but should not be too heavy, and a distinct contrast between the ground color and the lacing is necessary in order to bring out the lacing distinctly. The wing flights, secondaries, and the main tail feathers should be one uniform color. Likewise, the hackle, back, wing bows and saddle should be a uniform, clear blue-black. The tail coverts and sickles should be a dark blue, but not neces-

sarily quite as dark as the hackle, back, wing bows and
saddle. The fluff should also be laced, if possible to secure
it. It is of the utmost importance that the females for this
mating should be very clear and clean laced on the breast
and throat, and should have a good, dark hackle. It is not
necessary that the lacing on the back of the female be as
clear. The wing flights and main tail feathers should be
standard in color. Pullet mating—For this mating the
male should be of standard or exhibition color except on
the back, where more blue should show in the ground color.
Sickles that are considerably darker than the main tail are
desired, as they will give contrast to the tail color. The
females for this mating should be standard. It is very
important to have clear tails in both sexes. In either mating,
if it is necessary to use males which show white in face, they
should be mated to females which have very strong red or
even gypsy faces. In culling the chicks, it is necessary to
hold the blue birds until they throw their chick feathers,
which occurs at about three months of age, as it is impossible to tell anything about the lacing before this time,
since it does not show until then.

The following defects must be guarded against, in so far
as possible, in selecting the matings for this breed: too large
or too coarse combs; white in face of males; too large or
coarse individuals of either sex; arched back, often called
roached back; too sharp an angle at the juncture of tail and
back; birds of faded blue color; rusty brown top color;
unevenness of color in hackle, back, wing bows and saddles
of males; lack of contrast in ground color and lacing; weak,
washed-out color, especially in females; white under color
or white in blade of any feathers; white in flights or main
tail feathers (this is likely to be associated with white under
color and very light ground color); hens laying tinted eggs;
too low on legs; too high tail carriage, especially in males;
thumb-marked comb; lopped comb in males; too many
points to comb; Minorca comb in females; down between

toes; sickle feathers of males not darker than main tail feathers; undersized birds; stubs; occasional side sprigs; too short body; insufficient lacing.

The Ancona

This breed is very similar in type to the Leghorn, but has a distinct type of its own. It is about the same size or a trifle larger. The fowls are active and alert, rather well up on legs and with a smooth sweep of back and tail, showing a slight angle where these two sections join. The back is of good length and it slants downward slightly from the shoulders to tail, in this respect differing from the Leghorn.

The Single Comb Ancona

In mating this variety it is usual to use only a single or standard mating. Birds of both sexes are therefore selected which are as near standard as possible. The comb of the male should be neat and medium in size, strongly and firmly erect. It should be evenly serrated and have five points. While a male with a five-pointed comb should be selected for the mating, if it becomes necessary to choose between one with a four-point and one with a six-point comb, the former should be selected, as there is a tendency for the comb to have more rather than less than five points. There should be no tendency for the blade of the comb to follow the neck, but it should come straight out from the head. The front and first point of the female comb should be erect, and the rest turns or lops to one side. The comb in both sexes should be free from thumb marks.

Red sometimes occurs in the ear lobes, especially in males, and must be guarded against. However, birds with red in ear lobes rarely, if ever, produce offspring with white in face.

The black of the plumage should be free from purple barring. The white tip should be as pure white as possible, as there is a tendency toward an ashy hue in the white tip. The smaller and the whiter the tip in both sexes, the more desirable. There is also a tendency toward too much white

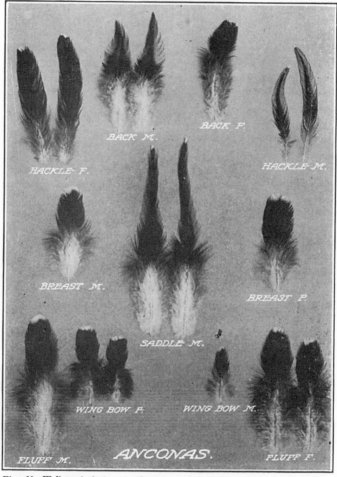

Fig. 68—Well-marked Ancona feathers. M indicates male and F female. (Photograph from the Bureau of Animal Industry, United States Department of Agriculture.)

in the wing flights and secondaries and in the main tail feathers. This must be selected against. In general, when a bird is fine colored otherwise, there is a tendency toward too much mottling. Then, too, there is a tendency for the birds to molt lighter with each successive molt, so that dark birds are preferable as breeders. See Fig. 69. There may also occur red in hackle, saddle and wing bows of males. Do not use such a bird for breeding except when the red feathers are few in number, and the male is out of birds free from this defect, and is an exceptionally good specimen otherwise.

Fig. 69—Single Comb Ancona female showing the tendency to grow lighter in color with age. (Photograph from the Bureau of Animal Industry, United States Department of Agriculture.)

When hatched, the chicks are orange and black. The young stock should not be culled too young, as many showing too much white will molt in good in color when fully developed. White in face is not a troublesome defect in this breed. There is a tendency on the part of many breeders to try to get their birds too large. Do not try to breed beyond standard size, as this tends to hurt their type and also their egg production.

In mating this variety, the following defects must be guarded against in so far as possible: too large combs in both sexes; side sprigs; too many points to comb; thumb-marked combs in both sexes; comb of female not erect in front, but flat on head, as in the Minorca female comb; white in face; red in ear lobes, especially in males; too much white in wing flights and secondaries and in main tail feathers; ashy color in the white tip in both sexes; red in hackle, saddle and wing bows of males; purple barring in both sexes; too high tails; willow legs which are more prevalent in males than in females.

The Rose Comb Ancona

In mating this variety, the same points must be considered as in the Single Comb variety, except in the matter of comb. The rose comb should be of the same shape as the Leghorn rose comb, but if anything slightly smaller in the male. It is necessary to guard against combs in both sexes which are too high or too broad, which have a spike with a tendency to follow the neck, or which show hollows either along the sides or on the top in the center.

CHAPTER VII

THE ENGLISH CLASS

The Dorking

There are three varieties of this breed, the White, the Silver Gray, and the Colored. In shape or type all these varieties are supposed to be identical. As a matter of fact, however, the White Dorking is decidedly inferior in type to either of the other varieties. The White variety also differs in that it has a rose comb, the other varieties being single combed. The three varieties differ in size, the Colored being largest and the White smallest.

In type, the Dorking is the ideal for meat production. The body of both sexes should be long, broad and deep, being markedly rectangular in shape. The body is also low set on short legs, giving the birds a massive appearance. The breast is very full and broad. The general slope of the back and of the body is slightly downward from front to back. In selecting breeders, choose those of both sexes which are longest of body and shortest of leg. Both these qualities are greatly to be desired and one is likely to go with the other. Select against small size of body, against short body and back, and against long legs. The rectangular shape of body and the rather angular junction of tail and body, particularly in the male, gives the birds somewhat of an angular appearance.

The neck of both sexes is short and thick, the male's being furnished with a full, long hackle which comes well around over the shoulders and the front of the neck.

The Dorking male's tail should be rather large, moderately low carried and fairly well spread. The female's tail, while carried at about the same angle as the male's, should be rather closely folded and not so widely spread as to have a fan-like appearance. In fact, it is really broader at the base than at the end. There should, however, be no appearance of scantiness to the tail, and the male should be well furnished with long sickles and an abundant and long saddle.

The single comb of the Silver Gray and the Colored Dorking is rather large and is erect in the male, but lops in the female. It carries six points or serrations. The rose comb of the White Dorking is a typical rose comb in both sexes, of rather large size.

The ear lobe should be red, but often shows some white or white tinge. A solid red lobe is preferred, other points being equal.

The Dorking has five toes. This characteristic distinguishes this breed from the majority of those commonly kept in the United States. The fifth toe should be close to and just above the usual fourth toe, but the two toes should be separated and show no tendency to grow together. It often tends to be short or misformed, but should be long and curve upward slightly. The other toes should be of good length and well spread. The shanks are short and stout, with good round bone. In color, both the skin and shanks are white.

The Dorking is a rather loose-feathered breed like the Brahma, but not so extreme in this respect as the Cochin. Tight feathering is a defect. The feathering is also of good length, as shown in the well-developed sickles, saddle and hackle of the male.

In order to keep up the size of the birds, it is important that they be hatched early and have an opportunity to get a long period of growth. It must also be remembered that the chicks will not stand damp, rainy weather very well.

In breeding the Dorkings the following defects must be guarded against in so far as possible: small size; short back or body; long legs; scanty appearance to tail; high tail; too large comb, either rose or single, especially in males;

prominent white ear lobes, especially in males; irregularly shaped or short fourth or fifth toe; off side spurs, that is, spurs on the outside of the legs in both sexes; short hackles and saddles in males, and too well-spread or fan-shaped tails in females.

The White Dorking

In breeding this variety it is common to use the single or standard mating. Select birds of both sexes which are as near standard as possible, both in type and color. Be sure that the combs of both male and female are set low and square on a good broad base. Do not use birds as breeders whose combs are unusually high or which tip or lean to one side. Use birds which are pure white in color, free from any brassiness or creaminess and from any foreign color or ticking. This variety is the smallest of the Dorkings and runs poorest in type. It is especially important therefore to guard against small birds and those with short body or too long legs. High tails are also rather troublesome. For defects to guard against which are common to the breed, see page 193.

The Silver Gray Dorking

This variety is medium in size between the smaller White variety and the larger Colored variety. In type, however, it runs quite good, probably as good as the Colored.

In breeding this variety a single or standard mating may be used, or a standard male may be mated to two kinds of females. This latter method is used by some of the most successful breeders. For this mating select a male which both in type and color is as near as possible to the standard. In type, select the largest body with the shortest legs. In color, a fully matured male should be as free as possible from any black on his white parts or any white on his black parts. He should be free from any frosting in breast or fluff. Often frosting shows in fluff when the breast is solid black. The white should be free from any rustiness or creaminess and the black should have a good greenish sheen.

Fig. 70—Well-marked Silver Gray Dorking feathers. M indicates male and F female. (Photograph from the Bureau of Animal Industry, United States Department of Agriculture.)

The hackle should be as clear a silvery white as can be secured, but usually there will be some striping in lower hackle. The clearer the hackle the freer the saddle will be from any tendency toward striping. The clearest top-colored males are quite likely to show frosting, however. With him mate two kinds of females, the first light females, which produce the best males, and the second dark females, which produce the best females.

The light females to select will be among the lightest of the flock, having an ashy gray top color, while the dark females best to use will be among the darkest of the flock, their general top color being a slaty or silvery gray. These dark females are standard in color. In selecting these two kinds of females, it will be found to be a great aid to hold two females side by side so as to compare the wings, back, breast, head, throat and hackle of each.

The lighter colored female should have ashy gray wings and back with fine stippling and as little shafting as possible. The breast should be as deep a salmon red as possible, free from mealy shading or edging of a lighter color. Mealy shading in the breast of females tends to produce frostiness in the breast and fluff of males. The head, throat and hackle should be silvery white, the upper part of the neck, the throat and the head should be nearly clear white and the remainder of the hackle as free from black as possible. Such females will produce males with hackles most nearly free from black.

The darker or standard colored female should have slaty or silvery gray wings and back, the stippling being extremely fine and as free from shafting as possible. The breast should be a clear bright cherry red, free from any mealy shading or edging of a lighter color. The head and throat should be silvery white, as entirely free from any brown as possible. The hackle should be silvery white, with a clear, distinct black stripe through the center of each feather.

Light shafting is present in practically all females and is

especially prevalent over the shoulders. Where it is present only in the shoulders, this is not a very bad defect.

In mating the Silver Gray Dorking, guard, in so far as possible, against the following defects, which are more or less common in this variety, in addition to the defects which are common to the breed (see page 193) ; stripy hackle or saddle in males; any white in the black or black in the white of males; rustiness or creaminess in the white, especially in the hackle; any very noticeable marking of brown on the head, wings or hackle of females; patchy or uneven color on breast of females; light shafting in females.

The Colored Dorking

This is the largest of the Dorkings and in type runs good. In breeding, it is usual to employ a single or standard mating. Both the males and females should be chosen as near standard as possible, both in type and color. In mating, it is necessary to guard against white, silver or silver white hackle in males, as this will lighten the color of the females. White or silver white hackles generally point to some previous introduction of Silver Gray blood. The hackle and saddle of males should match in color, but even when the hackle comes pretty good in color and is quite well striped, there is a tendency for the saddle to be silvery in color and to be lacking in striping. White in the wing flights of both sexes, but particularly in the male, must be guarded against, as must also too much silvery color in the wing bows and saddles of males.

It is also necessary to be careful not to use females which are in general too light in color. Females are desired whose hackles show as much white as can be obtained. At best this will only be a little white up on the head. Females should also be as free from gray as possible, and the lower part of the body should be dull black, while the upper part should be a blue black, with wide, prominent white shafting. For defects to guard against which are common to the breed, see page 193.

Fig. 71—Well-marked Colored Dorking feathers. M indicates male and F female. (Photograph from the Bureau of Animal Industry, United States Department of Agriculture.)

The Red Cap

This breed is one of good size, but not quite so large as the Silver Gray or the Colored Dorking. In type, it is somewhat more of an upstanding fowl, the shanks being medium in length rather than very short. The body is deep, broad and long, especially in the female, and the breast is very round and full. In general, the birds are shorter in body and more rounded in contour than the rectangular Dorking. The back in both sexes is straight and slopes downward slightly from shoulders to tail. The tail, which is well spread and of good size, is carried in a medium position, neither high nor low.

The comb is characteristic of the breed. In fact, it is the comb which gives the breed its name. It is rose, large in size, round, broad, setting squarely on the head, with no tendency to tilt or tip to one side, which may be so bad as to obstruct the sight. Like other rose combs, it should be free from hollows in the center or along the sides. It should be square in front and the spike should be well developed, of medium length, and extend straight back off the head, with no tendency to turn up or down. It is necessary to guard against narrow combs.

The ear lobe should be red. In this respect the breed is almost unique, for the eggs are white, and it is very rare to find a breed with red lobes which lays a white egg. White ear lobes or white in ear lobes not infrequently occurs and must be guarded against.

The feathering is fairly profuse, showing especially in long, flowing hackle and saddle in the male bird, and in a good-sized and well-furnished tail.

The legs and toes are a slate or leaden blue in color.

In mating, use a single or standard mating; that is, breeders of both sexes which are as near the standard as possible. In the male, be sure that the red sections of the hackle, back and saddle are a good red in color and are not light, approaching a straw color. This is important, as light colored males tend to produce too light or washed-out **colored females.** The black striping of hackle, back and

Fig. 72—Well-marked Red Cap feathers. M indicates male and F female. (Photograph from the Bureau of Animal Industry, United States Department of Agriculture.)

saddle should be a blue black or purple black, without any green sheen. The black and red wing color of the male should be distinct. The breast and fluff should be solid black. Red ticking or solid red feathers in the breast of males is a very bad fault. The general color of the red sections of the male should be rich and deep.

In the female, the ground color should be about the same as that of the Rhode Island Red, each feather terminating with a black marking shaped like a half moon. Shafting in all sections of the females is quite common and is a defect. Too light a ground color in both sexes, but particularly in the females, is quite a serious difficulty, and breeders must be selected which are free from this defect if possible. See Fig. 72.

In making the mating, the following defects must be guarded against in so far as possible: tipped or lopped combs; narrow combs; not a well-developed spike to comb; red ticking or solid red feathers in breast of male; shafting in all sections of females; too light a ground color, especially in females; too small birds; white in ear lobes; too light or straw-colored hackle, back or saddle in males; dull black stripe in hackle, back and saddle, lacking the purple sheen; black stripe in hackle, back and saddle, showing a green sheen.

The Orpington

The Orpington is a large bird of decidedly massive appearance. The very full, rounded breast and the slight forward tilt of the body of the most typical specimens gives the whole bird the appearance of tilting or tipping forward. The body is not only long and broad, but is very deep, and it is this depth of body which does much to give the Orpington its massive appearance. There should be no tendency toward a straight, flat-sided or narrow bird, as the Orpington should be round or well sprung over the back and sides.

These fowls are smooth and well rounded in all sections. Excessively large or too small birds should not be selected for breeders.

The back is really long, but because of its rise to join the tail and because of the extreme depth of the body, it appears, especially in males, as not very long, or even rather short. The length of back of the female is more apparent. The back has a slight slope downward from the tail to the shoulders. It is very broad as well as long.

The breast is well developed, being broad, deep and very full and rounded. It is therefore very prominent. Besides being long, broad and deep, the body has a long keel so that its underline is also long. It is fairly low set, but not so low that the feathering covers the hock joints.

The tail of both sexes is not carried high and is well spread. The junction of tail and back is perfectly smooth, so that there is no break or angle between the two sections.

The neck of both male and female is short and stout, which brings the head only slightly above the level of the tail.

The shanks are moderately short, but should not be too short. They should be of heavy bone and set well apart. A majority of males which have good size of body and of bone are inclined to be a trifle long in shank. The appearance of stubs and down in this breed is by no means rare and must always be selected against carefully.

The comb, which is single, and the wattles should be of medium size. Frequently, especially in males, they are too large. While the comb may be and should be slightly larger than the typical Plymouth Rock comb, it should not be excessive. It should have a firm base, should be well set on head and have five even and symmetrical serrations. The blade should follow the shape of the head and neck rather closely. The female comb should be proportionately smaller, so that it will not twist or lop. Thumb marks in males and side sprigs in both sexes are to be avoided.

The ear lobes should be a good solid red, entirely free from any white. White in ear lobe, as in the Plymouth Rock (see page 71), is sometimes troublesome, but this defect is more apt to develop with age.

The feathering should be rather loose and abundant, but should not be of such excessive length as to approach the Cochin.

In breeding the Orpington, the following defects common to the breed must be guarded against in so far as possible: too large or too small birds; deficient breast; short back or body; body lacking in depth; narrow back and body; body not set low enough, or too long shanks; too fine or light bone of shanks; too short shanks; too large comb; blade of comb not following the head and neck; twisted or lopped comb in females; thumb-marked comb; side sprigs; stubs and down; white in ear lobe; too long or too loose feathering; split tail in males; too long and narrow heads; straight, flat-sided birds; too light eyes.

The Buff Orpington

In mating this variety, it is most common to use a single or standard mating, in which breeders of both sexes are chosen which most nearly approach the standard requirements. The male should be selected which has the evenest and the soundest color throughout, that is, which is free from white, black or black peppering. The exact shade of color is not so important as to have the color even, so that the different sections will harmonize and blend with one another without marked contrast. His under color should be as sound as possible and particular attention should be paid to soundness of under color in hackle and at base of tail and to the wing quills, to see that they are sound buff, not white. A male with good surface color, but with a weakness or even white in under color, may produce many fine-colored pullets, but will rarely produce a cockerel sound in his wings. The cock bird which molts in solid in under color of hackle and·saddle is more valuable as a breeder than one which is sound as a cockerel but fails to molt in sound. Most males carry a little smoke or brownish cast or black peppering in the main tail feathers. This will only be noticed when the bird is handled. Such a male is better for breeding than one with a sound buff tail, but with weakness of color in other sections. A little smoke in the short wing coverts or in the secondaries is not a fatal weakness in a breeding male, but one free from this, if otherwise good, is preferred. In general, in the matings, if the male is a trifle light, use females one or two shades darker, and if he is a little dark, use females one or two shades lighter. Do not under any circumstances make matings in which the birds are the extreme in color, as the majority of chicks will be mottled and mealy. It is especially important to select against females as breeders which show mealiness on the wing bows. The use of such females will lead to general color troubles in the offspring, such as white in the secondaries of males. White in the flights of males may only mean that there has been some accident or check in the growth of the individual concerned, and if that is the case this is, of course, not a serious breeding defect. If white occurs in the secondaries, however, it is a serious defect, which usually comes from the use of mealy-winged females as breeders. Mate birds which are not more than one or two shades of color apart.

It is a common error to leave out of all these matings any birds with real strength of color pigment. This leads to a loss of color. Few sound-colored cockerels will be produced from a light mating, and while the females will be better in color, the majority of pullets will molt too light as hens. Birds of stronger color must be bred carefully and perhaps only occasionally, but they must not be discarded entirely, or the flock will lose in color. In general, the same color considerations in mating this variety apply as in other buff varieties. (See page 89.)

Red on the outside of the legs of males should not be considered a defect. Most birds showing this are very vigorous.

The following defects, in addition to those common to the breed (page 203), must be guarded against in so far as possible: uneven color; too heavy color on shoulders and back of males; black, white, smokiness or black ticking in wings and tail; too light or white under color at base of hackle, at base of tail or in saddle of males; mealiness in females and too light color in females.

The Black Orpington

This variety is the most massive, the lowest set, and the most loosely feathered of the Orpingtons. It is generally considered to run the best in type. In mating this variety, the same general color considerations must be observed as in any other black variety. The single or standard mating is generally used. Select a male with very dark under color and with as good surface color as possible. It is rare to find the bird with the best surface color, strongest in under color. One of the most common defects in the surface color of males is purple barring. It is also one of the most difficult to eradicate. Red in hackle, saddle and back is also a somewhat common defect. Mate with the male selected to head the pen, females with as soft a color as can be had with dark under color. The growing stock and the old stock during the molt must have good care and feeding or their color will not be good, no matter what the breeding.

Birds with black beaks, having as few white markings as possible, should be used. The darker the eyes the better, since a black eye is desired, but nearly all eyes have a dark brown cast. Light eyes should be selected against.

The White Orpington

In breeding this variety, it is usual to employ a single or standard mating, selecting breeders of both sexes which most nearly approach the standard both in type and color. The same general color considerations apply to this variety as to any other white variety. Birds of both sexes should be as white as possible, free from brassiness, creaminess, and from any black ticking or foreign color. Brassiness is quite prevalent and shows both as a brassy tinge to the top color of the male and as a brassy tinge on the hackle of the females. While creaminess is also rather troublesome, it may often be due to the sappy or immature condition of the feathers of young birds, or birds which have just molted. If that is the case, this creaminess will be lost as the feathers mature, and is therefore not a defect which need count in breeding. In selecting breeders, great care should be taken to see that the quills are white. If the quills are yellow, this is apt to be reproduced in the offspring. Black ticking throughout the plumage, but particularly in the wing and tail, is quite common, especially in females, and is, of course, undesirable. Black ticking in males is mostly found in the tail. However, it frequently happens that birds showing some ticking are whitest in color. Foreign color in the plumage may occur as red, buff or partly black feathers, especially on the shoulders of males. This should be selected against, as it is apt to be transmitted to the offspring. A little red occasionally occurs in the hackle or saddle of males. While this is undesirable, a male possessing it, if an exceptional individual otherwise, may be used. as it does not appear to be transmitted to a very large proportion of the offspring. In selecting good color, it must be borne in mind that the very late hatched cockerels are always whiter than the earlier hatched cockerels. This is simply due to the fact that the mature plumage of the late hatched birds has not been so long subjected to the effects of the weather as in the case of the earlier hatched birds. The whiter color, due to later hatching, only lasts therefore through the first molt and must not be given undue weight in selecting breeders. The shanks in this variety are sometimes blue or creamy. Both shades of color are defects and should be selected against, as a white shank is desired. Creamy bills or beaks and too light eyes are other defects to be guarded against. For defects common to the breed, which must be guarded against in so far as possible, see page 203. Some of the newly hatched chicks are gray or smoky in color and these are apt to develop into the whitest birds.

The Blue Orpington

In breeding this variety, both the single or standard mating and a double mating can be employed. In the single mating, use a dark male in which the ground color is dark blue and the hackle, back, wing bows and saddle are of such a dark blue as to be almost black. This male should be well laced on throat and breast. With him mate a well-laced, standard-colored female. It is from the dark male that the good lacing on the female offspring is obtained. If a light male is used there will be little or no lacing on the female. Some of the females produced from this mating will be black. Keep these and breed them to the dark males in order to get males with good dark color. A blue male, mated to a black female, will get as large a percentage of blue chickens as will be obtained in a straight blue mating. Do not use males as breeders which show brassy, red or straw-colored head or back. A little light in under color is not objectionable in the male. Do not use females with light hackles, as this causes poor ground color and may result in obtaining white flights.

The general color of the Blue Orpingtons is the same as that of the Blue Andalusian, except that the males tend to run darker in color and the lacing on both males and females is not as good as in the Andalusian. Sometimes Blue Orpington chicks which have black plumage while they are still in their chick feathers will molt to a blue. This never occurs in Blue Andalusians.

Since the same principles apply in mating the Blue Orpington as in the Blue Andalusian, the reader is referred to the matter on Blue Andalusians (page 181) for more detailed information concerning the matings for blue color.

The Cornish

The Cornish is essentially a meat type. With its typical erect carriage, its appearance of great strength, its heavy flesh, good vigor, its low set body, its extreme breadth, the development of its breast and its close feathering, it is quite distinct from any other breed. The general outline, as viewed from the side, may be likened to an egg with its larger end, represented by the upper breast, foremost and uppermost. The erectness of carriage is sufficient to bring the front point of the breast bone in line or on a level with the junction of the tail and the back.

The body is low set, due mainly to the shortness of the shanks and to the fact that the thigh is set high up on the body under the wing. Old birds should be very low, broad and deep, while young birds, unless up to adult weight, are apt to appear a little taller or more leggy and not quite so blocky. However, if this deficiency is slight, they will usually gain the desired type as they settle down and fill out with age. In selecting low-set birds for breeders, avoid the pit game type. The body should be short or medium in length, not long. The body should be well rounded at the sides, the ribs being well sprung. Never use a bird with a flat-sided body.

The back should have a decided slope downward from the base of the neck to the tail. The backbone should be free from angular joints and nearly straight, but the feathers should lie so tight as to make the top line of the back appear slightly convex. The back should be very broad at the shoulders and this breadth should be carried back well to the hips, from which point it narrows to the stern. The breadth between the hips is necessary to make room for the broad breast, carried well back between the legs. The shoulders are very broad and heavy and are carried level or slightly drooping to conform to the slight slope of the back from the spine toward each side. While the back has a decided

downward slope, the angle between the back and the tail is not sharp, due to the low carriage of the tail, but it is distinct. There should be no cushion in the female.

The wings should be short, well rounded, tightly folded and carried close to the body. The great breadth of shoulders is desirable if for no other reason than because it means strong wing muscles,

Fig. 73—Dark Cornish male showing massive compact low set body, broad shoulders, broad chest, well-spread legs set in line with the shoulders and fearless expression. (Photograph from the Bureau of Animal Industry, United States Department of Agriculture.)

which in turn require strong breast muscles, thus making for breadth and roundness of breast.

The breast is unusually broad, prominent and well rounded, and extends well back between the legs. It is this breast development which makes this breed so desirable for the table.

The neck is rather short and very strong, with the end of the hackle just touching the shoulders. It is nicely arched just below the head, but the rest of the neck is straight and comes straight off the body, having no suggestion of the curved or goose neck. The neck, particularly in front, is not full feathered. Avoid a wide, full-flowing cape, as it detracts from the broad shoulders.

The tail should be small, short, narrow and closely folded. The sickles and coverts should be narrow and just long enough to cover the main tail feathers. The tail should be low carried, either horizontal or slightly drooping. Long, bushy tails, which may occur especially in the White Laced Red Cornish, should be avoided. Never use a male as a breeder which has long, loose or abundant saddle feathers, as he is apt to produce females with cushions.

The legs should be wide apart and straight, with no suggestion of knock-knees or bent or cow hocks. They should be set well forward under the body in line with the shoulders, giving the bird a well-balanced appearance. The thighs should be of moderate length, not short, and should be thick and bulging, with firm muscle. The shanks should be large and short, denoting strength and causing the low-set body appearance. The bone of the legs should be heavy, showing plenty of substance, well rounded and the scales smooth. Avoid light or flat bones in the shanks. Discard as a breeder any birds with stubs. The color of shanks is yellow or orange. The feet should be large and sound, the toes well spread, rather long and very strong. See Fig. 73.

The head should be short, deep and broad, with a short, strong, well-curved beak. A narrow, pinched head is undesirable. The eyes are set wide apart, with the crown or brows projecting somewhat over them. This projection of the crown is generally thought to indicate strong constitution. It also gives the bird a rather fierce, savage expression. The eyes should be rather large, bright and prominent. They are pale yellow, approaching pearl in color. Discard any birds as breeders which have red eyes, as these have been difficult to breed out of the Cornish.

The comb is a distinct pea, like the Brahma, but is rather small and fits the head closely. Single combs but seldom occur and must, of course, never be used for breeding. Too large and tipped or lopped pea combs must also be avoided. The wattles should be small. Absence of wattles is preferable to large wattles. The ear lobe should be small and red. White in the ear lobe must be avoided.

The feathering is characteristic of the breed. The feathers must be short, narrow and wiry, full of life and very glossy. When the feathers are bent up from the body they should snap back into place like a spring. The bird's plumage should be almost skin tight, fitting like a coat of mail. Any tendency toward looseness of feathers, softness, or length of feather must be carefully selected against.

In size, the Cornish is large and very meaty. The White and White Laced Red varieties are smaller than the Dark, although the standard weights for the Dark and White are the same. In mating, it is well to use a large cockerel of good type, say one of 10 pounds, if possible, in the Dark, and one of 8 pounds in the White and White Laced Red varieties, with hens two or three years old, and of large size, strong bone, and great substance and type. Select a cockerel very wide between the thighs and with short, massive shanks. However, do not run to extremes, as the Cornish cockerels, which are very low set, never make the size of those which at first appear to be a little higher on legs. If cocks are used, select for type, bone, vigor and activity, rather than extraordinary size, as the very large cocks are apt to be clumsy and their fertility much lower than a moderate sized, vigorous male.

In mating, avoid in so far as possible the following defects which are characteristic of the breed: too leggy; pit game type; long body; narrow back and body; narrow back between hips; cushion in the female; curved or goose neck; too open tail; too high tail; long, bushy tail; long, loose or abundant saddle feathers; knock-knees; cow hocks; short thighs; too long shanks; too light bone of shank; flat bone of shanks; stubs; red eye; single comb; too large comb; tipped or lopped comb; large wattles; white in ear lobe; general looseness, softness or length of feathering; too small size; lack of vigor; flat-sided body; narrow, pinched head, and wide, full flowing cape.

The Dark Cornish

In mating this variety, it is most usual to employ a single or standard mating. Select breeders of both sexes which approach the standard as nearly as possible, both in color and type, being sure that they are birds of great vigor and substance. Also select, if possible, a male of good color which is out of a well-marked female. In such a mating the principal defects to guard against are white in the under color of both sexes. This is especially likely to occur in the hackle, back and saddle of males. White in the primary and secondary wing feathers must also be avoided. Where this is present in only one wing it is apt to be due to injury, and is not serious, but where it is present in both wings it is apt to be in the blood and is more serious. Also select a male whose breast is black, showing no lacing of red.

In the female there is often an uncertain conception of what constitutes proper color. Females are shown ranging in ground color from a washed-out lemon to a dark, smutty walnut. The ideal ground color of the females

is a clear, dark red, like old polished mahogany, with a
double penciling of lustrous, greenish black. The first
penciling is in reality a narrow lacing running evenly

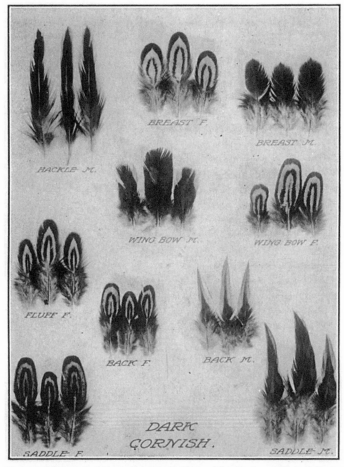

Fig. 74—Well-marked Dark Cornish feathers. M indicates male and F female. (Photograph from the Bureau of Animal Industry, United States Department of Agriculture.)

around the feather. The second penciling is within the
ground color of the feather, which it divides into two
nearly equal parts, the central part being slightly wider
than the outer part. This second penciling is also narrow
and is crescentic in form, running practically parallel to
the outer penciling or lacing. The contrast between the
black penciling and the red ground color should be distinct, and heavy or wide penciling, which tends to cover
too much of the ground color, is not desired. Heavy
tips of black at the ends of the feathers are likewise undesirable. Sometimes pullets occur in which the black is
too predominating, but in the second year, after the molt,
this may be corrected. Females having only a single
penciling or lacing and those which are triple penciled
occur and should not be used unless special matings are
made. Pullets which are very tight and hard feathered
sometimes appear to be single laced, but when they are
handled and the feathers opened up they are found to be
really double laced or penciled. The females should be
well laced or penciled out into the tail, while stippling in
the tail coverts is undesirable.

The following matings are advised by another breeder:
The best mating for standard color is to select double
penciled females as nearly ideal in color, markings and
type as possible, and mate to a male as nearly standard
as possible out of the same kind of a hen.

When the mothers of the available males of the desired color are not known and it is feared that they may
be irregularly marked or too dark females, which not infrequently produce good males, it is better not to use
such males, as the females from him would not be likely
to be good. Instead, choose one with the shafts of each
feather in the under color of breast, body, back and wing
bows, red or black heavily striped with red. Such males
usually have a well-striped hackle, and the red shaft in
the hackle is quite as important in producing good fe-

males as good males. The red shaft helps to ward off triple penciled and too dark females, but does not need to extend out into the black surface color of the breast and body, thereby spoiling the exhibition color of the bird by being likely to show splashing in the breast. Such a male would be best for a general mating from which both males and females of good color are desired. If the greatest possible percentage of exhibition females is desired, a male should be used which shows in addition a little red lacing on the surface of breast and body. Owing to the general tendency of the variety to go darker, some exhibition males might also be obtained from such a male.

Owing to this same tendency for birds of the variety to darken in successive generations, it is possible to breed single laced hens to quite dark males and produce some very fine double laced or penciled females.

A red shaft in the under color of the body plumage of the females the same as for the male is preferred by some breeders. A little wider bay may be allowed in the hackle of the female than in the male, but in neither sex should the red be allowed to creep to the tips or the edge of the feathers. If it does, it is likely to prove to be a forerunner of the same trouble in certain parts or all of the body plumage. The black in the top color of the male should be a greenish black and should be free from any purple cast or barring. Birds with a little white in under color are more apt to throw birds free from purple. Males with this greenish black in hackle and free from purple will help the lacings or pencilings throughout the female's entire plumage. Purple or purple barring is hard to overcome. Some females retain their rich ground color with successive molts, but many do not, coming lighter, often a faded or pale bay after the molt. While the females which hold their color are the most valuable for breeders, the loss of color due to age should not be allowed to count too heavily against females which were good as pullets. The under color of both sexes should be dark slate. It may become a little lighter with each successive molt, especially in males, so that a lighter under color in an old male, which was good as a cockerel, is not a serious breeding fault. The feathers, especially over the back and saddle, should be rather narrow. See Fig. 74.

In mating this variety, the following defects in addition to those common to the breed (page 212) must be guarded against in so far as possible: white in under color of both sexes, but especially in hackle, back and saddle of males; white in wing primaries and secondaries, especially in males; laced breast in males except for special matings; triple laced or penciled females; single laced or penciled females, except for special matings; lacing or penciling not carried well out into tail of females; stippling in tail coverts of females; too light or dark, smutty ground color in females; too heavy or wide, black penciling in females; heavy black tips to feathers in females; red in hackle of either male or female extending to the tip or edge of the feathers; purple cast or barring in the black of the male's top color; any indication of cushion in females or long saddle feathers in males.

The White Cornish

This variety is identical with the Dark Cornish, except in the matter of color. While the standard weights for the White and the Dark Cornish are the same, as a matter of fact, the White variety runs smaller than the Dark. A single or standard mating is universally used. Both males and females are selected which approach the standard requirements as nearly as possible, both in type and color. They should be as white as possible and as free from any foreign color as can be secured. Breed for a brilliant, sparkling live white, rather than for a dull, lusterless, dead white. Black ticking sometimes occurs and should be selected against. Occasionally a portion of a body feather or two may be greenish black. This is not a sign of impurity of breeding, nor is it, if not frequent in appearance, a serious breeding defect. Very much red or a great deal of gray in the plumage must be carefully avoided. Perhaps one of the most serious defects is brassiness. This shows most commonly in the hackle, back, wing bows and saddle of males, and consists of a yellow or brassy color on the surface of the white. It can be bred out by careful selection, and the choice of breeders showing white quills and white shafts to the feathers will be of assistance in this effort. Brassy birds are in general less likely to show black ticking than pure white birds. Creaminess which shows as a slight yellow tinge in the under color and quills is less troublesome. It is often due to the sappy condition of the feathers after the molt and usually disappears as the feathers mature or soon after the molt is completed.

In mating this variety, the following defects in addition to those common to the breed (see page 212) must be guarded against in so far as possible: black, black ticking, red, gray and brassiness.

The White Laced Red Cornish

This is the most recent variety of Cornish, and in the effort to fix color, type has been more or less neglected, with the result that this variety is smaller in size and not so good in type as the Dark. In mating, it is usual to employ a single or standard mating. Birds of both sexes are selected which are as well laced throughout as possible, that is, the lacing narrow, the white free from any foreign color, and the red a good, rich red, the contrast between the white and red being clear and distinct, with no sign of mossiness.   In general, the birds vary in their degree of color, depending upon the width of the white lacing. If this is wide, the bird is light, and if it is narrow the bird is dark. The lacing on the breasts and bodies of the males is likely to be a little heavier or broader than that of the females. However, the males are almost as well marked as the females. The hackle feathers in both sexes should have a broad stripe of dark red in the center, leaving only a narrow, white lacing, but the red should not extend to the edge or tip of the feathers. If it does, the same trouble may be expected in the rest of the plumage. The color of the red of the male should be even in hackle, back, wing bows and saddle, but not infrequently un-evenness occurs. Another defect in males is white on the outside of the secondary wing feathers. These should be red, except for the white lacing. Sometimes, also, lacing is absent on the back and saddle, especially of uneven colored, dark red males.

In the female good breast lacing is perhaps the hardest to obtain and most often lacking. Females with lacing of medium width as pullets are generally laced in every section, and will molt into hen plumage having the pullet color markings. Females with white hackles also occur and on these birds the breast lacing is often good. The under color of both sexes should be white and a blue cast to the under color of females must be guarded against.

While line-bred specimens of both sexes as near standard as possible will produce good birds of both sexes, there is a tendency for the color to be lost to some extent after a time. This makes it desirable to use what may be called a necessity mating to keep up color. Such a mating consists of mating light colored or wide-laced females with a standard male of very red color and very narrow lacing. This mating will produce both standard males and females, but not in as great a proportion as the straight standard mating. One of these standard offspring, usually a male, is then bred into the old line to strengthen color. The reverse of this mating, or light males with standard females, can also be used for the same purpose.

Fig. 75—Well-marked White Laced Red Cornish feathers. M indicates male and F female. (Photograph from the Bureau of Animal Industry, United States Department of Agriculture.)

As chicks, the White Laced Red Cornish come almost white, and the first feathers will be white until they are six weeks old. Pullets have been known to remain white until three or four months old before a red feather showed. It is well, therefore, to carry chicks at least four months before culling, to see how they are going to molt in.

Males never change in the adult plumage, but females of the well-laced neck and back often become somewhat mottled in appearance and remain so until early spring, when they again show the lacing of the pullet. The females of the white-necked type will come into hen plumage just as they were as pullets.

In mating this variety, the following defects in addition to those common to the breed (see page 212) must be guarded against in so far as possible: too small or too leggy birds; too wide lacing; lack of distinct contrast between the red and white or mossiness; red running out to the edge or tip of feathers, especially in hackle of both sexes; unevenness in red color in hackle, back, wing bows and saddle of males; white on outside of secondary wing feathers in males; absence of lacing on back and saddle of males; poor breast lacing in females; white hackles in females, and a blue cast to the undercolor of females

The Sussex

While the Sussex is a large breed, having a weight equal to that of the Dorking, it does not give quite the same appearance of massiveness. This is probably due to the fact that it is a little shorter in body and is a little higher set on legs. It is much on the Dorking type, however, having a general rectangular shape, as viewed from the side. It is deep bodied and carries a good deal of flesh. The back should be fairly long, although not so long as the Dorking back. There is somewhat of a tendency for the back to be too short, thereby approaching the Orpington type, especially in the Speckled variety, and this must be guarded against. The back is flat and wide across the shoulders, but tapers down somewhat toward the tail. The slope of the back is downward from the shoulders to the tail.

The breast should be very prominent and well developed, being carried well forward. It should also be well rounded, and a flat breast should be selected against.

The tail is well spread and is carried fairly low, about the same as that of the Dorking in the female, but a trifle higher in the male. A high tail is not uncommon and must be selected against. The tail is well spread, more so than the Dorking, but is not so long. This difference in length is especially marked in the female. The Sussex tail is not furnished with either sickles or saddle feathers which are as long as those of the Dorking. The neck should be medium in length in both sexes, being a little longer than the Dorking, while the hackle should be fairly full, but not so full as in the Dorking.

The legs are comparatively short and stout and are set wide apart. They are longer both in thigh and shank than the Dorking, however, and it is for this reason that this breed is not so low set. The shanks should be white in color and the toes four in number. Black spots on the shanks must be avoided in breeders, as must also the fifth toe, which occasionally appears in the Light variety. The shanks should also be free from stubs. The ear lobe should be red in all varieties. White sometimes occurs and should be selected against in the breeders. As in other red ear lobed varieties, paleness or the white which develops with age is not so severe a breeding defect as the white which occurs in young stock. (See page 71.)

There is a tendency for the eyes to come too light in color and this must be looked out for. A red or reddish **bay eye is desired.**

The character of feathering in this breed is a little closer, that is, not so heavy, as in the Dorking.

In breeding the Sussex, the following defects more or less common to the breed must be guarded against in so far as possible: too short back; flat breast; too high tail; too large comb; too small comb; too many points to comb; lopped or badly twisted combs in females; side sprigs; black spots on shanks; stubs; fifth toe, especially in the Light Sussex; white in ear lobe; eyes too light in color.

The Speckled Sussex

This variety as bred by the foremost breeders and as winning under the judges in the shows, is slightly different from what the standard calls for. The standard calls for slate under color, shading into white at the skin, while breeders are working for a slate under color, shading into pink or salmon. In reality, the Speckled Sussex is a Red Sussex in which the feathers end in a black bar with a white tip, and in the male with a black stripe in the lower end of hackle and saddle feathers.

In mating, both the single or standard and the double matings are used. In the single mating, birds of both sexes are chosen which most nearly approach the standard, both in color and type. The under color should be slate, shading into pink or salmon. If there is too much dark in under color, this tends to produce a heavy, black bar in the breast of males, which will cause the breast to appear too black. White not infrequently occurs at the base of hackle and saddle of males, and is not desirable. Males are preferred as breeders which have black and white main tail feathers. There is a tendency and a desire to do away gradually with the white main tail feathers. However, if the main tail feathers are solid black, except for the white tip, they are apt to cause too dark breast in males and too much black in the back of females. On the other hand, too much white in main tail feathers is likely to cause the plumage to run too light in the offspring. White flights in males are

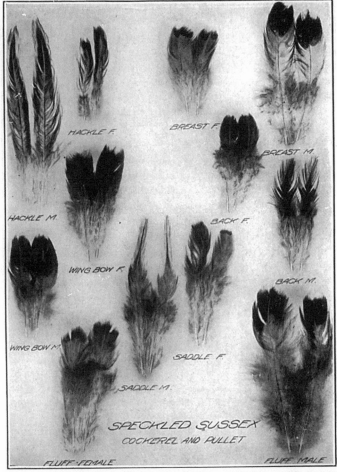

Fig. 76—Well-marked Speckled Sussex feathers. M indicates male and F female. (Photograph from the Bureau of Animal Industry, United States Department of Agriculture.)

undesirable as well as solid white feathers in any part of the plumage or solid black feathers in hackle or saddle. In the male there may be too large or too small a white speckling in the breast or too large or too small a black bar in the same section. Too large white tips will cause the breast to be too white, and too large black bars will make it too dark in appearance. The black bar which separates the white tip from the ground color of the feather should be very clearly defined, with no tendency for the black and white or the black and red to mix. The breast should not appear wholly black and white, as some of the red ground color of the feathers should also show. The tail coverts also have a tendency to carry too much white, but white tips are desirable. Feathers lacking the white tip in hackle and saddle of males must be guarded against, as this will result in the feathers being black tipped. Purple barring in the black of the tail must also be avoided. The striping in hackle and saddle of a standard colored male should be definite and distinct, but the tip of the feathers should be white. There should be a sharp contrast between the black bar and the white tip, and between the ground color and the black bar, and none of these colors should run into another. Sometimes weakness in color in females is shown by the black bar being absent so that the red runs up to the white tip. The female may also have defects in the way of white main tail feathers, black hackle, solid white feathers in wings and a mossiness or stippling of black in the red ground color.

In the female, there is also a tendency for the hackle to grow lighter with age, and this must be considered in breeding. A hen that holds a pullet colored hackle is considered extra good in that section, and especially valuable as a breeder. Shafting must also be avoided in the females.

Where the double mating is employed, one breeder describes the matings as follows:

Cockerel mating.—Select a standard male of brilliant color and with definite striping and speckling in hackle and saddle. The male should not be extremely heavily striped, but rather a medium striped bird. Do not use a male with a lemon colored back. The tail covert tips of this male must be white. With this male use females of standard type, which may be coarser marked than the standard colored females, that is, well but not too heavily speckled, and the speckles should be pure white. They should have a definite, clear striping in hackle. The wings and tail should be dark, that is, should have but little white. In general, the females used in this mating are rather dark in color.

Pullet mating.—Winning males of this variety have too long and heavy a black stripe in saddle to get good females, as such males produce females whose ground color is stippled with black. The male used should, therefore, have a definite hackle striping which follows the outline of the feathers, but is not too heavy, and the tips of these feathers must be white. The saddle should have only a small, black stripe, which should break, that is, not run down into the under color. In general, this male should be richer, that is, slightly darker in ground color than the standard colored male. It is important that the ground color of the male should be very prominent and striking. The breast should be clear and distinctly marked, while the speckling in the saddle and neck should be rather fine. The females to use should be the best exhibition females available. They should have clear ground color, especially in back, and wings and breast free from mossiness. The speckling should be clear white, and the main tail feathers black ending in white tips.

Another breeder briefly describes his method of double mating as follows:

Cockerel mating.—Use standard colored birds of both sexes, bearing in mind that breeders desire in exhibition males a lustrous, deep mahogany red.

Pullet mating.—Use the deepest red male which can be obtained and mate with standard colored females. In any mating, keep away from any stippling in females and from a light, cherry red or lemon color in males.

To summarize briefly, the following defects in addition to those common to the breed (see page 222) must be guarded against in mating this variety in so far as possible: wholly white main tail feathers in both sexes; white at base of hackle and saddle of males; lemon colored or too light back in males; striping not definite in hackle and saddle of males; tail coverts lacking white tips in males; too large white speckling causing too white an appearance to breast of males; too large black bar causing too dark an appearance to breast of males; no red ground color showing in breast of males; black tips to hackle and saddle feathers of males; tail coverts showing too much white in males; black hackle in females; solid white feathers in wings; mossiness; stippling or peppering of black in the ground color of females; black bar absent in females; black mixing with white in the tip in females; purple barring in the black of males' tails; female hackle which gets too light with age; shafting in females; solid black or white feathers in plumage of both sexes.

The Red Sussex

The single or standard mating is most commonly used in breeding this variety, as high class exhibition specimens of both sexes can be produced from such a mating. Some breeders, however, employ a double mating in order to produce especially fine specimens. In the Red Sussex there is a natural tendency for the males to come darker than the females. There is also a decided tendency for the ground color of the female, especially in back and breast, to be ticked, peppered or penciled with black. This is very undesirable and must be guarded against.

The single mating consists of a male which should be standard in color, having the dark mahogany red in all sections, as called for by the Standard, and with slate under color, shading into red at the base of the feathers. Light colored hackles are a common defect and should be avoided, an even shade of top color in hackle, back and saddle being desired. The females should also be standard in color and if the ground color of the females is selected to match that of the male's breast, it will be found to be a good guide. The females should also have a clear ground color, free from ticking, peppering or penciling of black, especially in back and breast. Be careful, also, if the breast is clear that it is not too light red in color, an even shade of surface or top color being desired, as in the male. The female should also be darker in under color than the male in order to produce the rich, dark surface color in the males. Be careful, also, that the male in this mating shows no white at the base of hackle or saddle, which is more troublesome and frequent at the base of the hackle. Black splashes on the male's breast and black in the wing bar of the male are likewise defects which should not be used in a standard mating as above described. The wing flights and secondaries of both sexes should be free from peppering.

Where the double mating is used, the following matings are made:

Cockerel mating.—Select a male of standard color, except for the under color, which preferably should be red. The use of a male with slate under color will result in a large number of cockerels showing black in breast and wing bar. Use females which are strong in color, being mahogany in hackle and free from any orange tinge. If the females show ticking in the back, this need give no concern. The under color of the females should be slate.

Pullet mating.—Use a male with an excess of color, showing strong under color, the slate running to the skin and the only red in under color being in the quill. This male should also show good black points, the wing showing black markings like that of the standard Light Brahma. The hackle should be practically free from ticking. With this male, mate good, standard colored females which have clean, red backs and wing bows free from ticking or penciling. The under color of the females should be red, free from slate, but birds showing a slight bar of slate may be used. From this mating should come good females, with clear backs, but some of the males will have black breasts.

In mating Red Sussex, the following defects in addition to those common to the breed (see page 222) must be guarded against in so far as possible: black ticking; peppering or penciling on ground color of females, especially on back and breast; too light red breast in females; white at base of hackle in males; black splashes on breast of males; black in wing bar of males; white in under color in both sexes; peppering of black in wing flights and secondaries; light or orange hackle; white on wings; white on tips of feathers.

CHAPTER VIII

THE POLISH CLASS

The Polish

The Polish is a breed which in type and size is much like the Leghorn. The size is, however, rather variable, but the birds should be at least as large as Leghorns, and some specimens are produced which are considerably larger. The Polish is a neat, trim-bodied bird and should be set fairly well up on legs, but there is some tendency to be too low set. The body is long, the breast fairly prominent and well rounded. The body has a distinct slant downward from front to back, and the back line shows this also. The tail, which is long and well spread, and in the male profusely furnished, is carried low, somewhat more so than the Leghorn.

The legs and toes should be blue or a slaty or leaden blue in all Polish, and white legs must be avoided. There is a tendency for the legs to fade with age toward white. The shanks and toes should be clean, but stubs and down sometimes occur, and birds showing stubs should be discarded for breeding. The larger the crest and the older the birds, the greater is the tendency for down between the toes.

The comb is V-shaped and should be small, as the smaller the comb the better it is liked. It is, however, apt to come too coarse, especially in the males, and sometimes approaches the leaf type of comb. A natural absence of comb is very desirable. It is important not to breed from birds which have closed nostrils, that is, the low, slit-like nostrils occurring in most breeds, as such birds are apt to throw single combs, which would disqualify the specimen. The breeders should show open or raised nostrils. By open nostril is meant a raised nostril, the opening of which is more circular in outline than the low or so-called closed nostril, and which causes that portion of the upper bill above it to be raised or arched. See Fig. 88.

The crest is undoubtedly the most characteristic feature of the Polish. It should be large and full, and in mature

Fig. 77—Bearded Golden Polish male showing lightly feathered or whisk broom-like crest, which is undesirable. (Photograph from the Bureau of Animal Industry, United States Department of Agriculture.)

males the crest feathers are long and should follow the neck and fall down well over the sides of the head. Avoid any tendency for the feathers in front to twist and fall forward over the bill. A breeder should not be used whose crest splits or parts in the center. Both males and females some- times show a lightly feathered, whiskbroom-like crest, which is undesirable. See Fig. 77. The crest should be straight on the head, not crooked or leaning to one side. Never breed from a bird which shows a crooked crest, as this is very apt to be reproduced and is difficult to get rid of. In

examining the crest to see whether it is good, feel of it
to see that it sets solidly on the head and cannot fall to
one side. Many breeders feel that it is fairly easy to pick
out the birds which will have the best crests when they are
first hatched. Such chicks will show larger, fuller crowns
and the best crests will be round and flat on the chicks, while
the poorer, lighter-feathered crests will show as small and
peaked.

In mature birds, the crest is sometimes so large that it
falls down over the eyes, obstructing the sight, in some
cases eventually causing the bird to go blind. Such a crest,
which interferes with the bird's sight, prevents them from
eating freely or breeding well. It is therefore frequently
necessary to clip or trim the crest fairly closely in preparation for the breeding season. Instead of clipping, a rubber
band can be placed around the crest to hold the feathers up
out of the eyes, but this is not as good practice, because the
crest feathers will hold the water if they get wet, and this
may lead to colds and roup. It is also quite common practice to pluck or pull the crest to increase its size. This is
done gradually, being sure that the feathers are dead before
they are pulled. If they are not dead, injury is likely to be
caused to the feather follicles and the new feathers are apt
to come in white, which would, of course, be very undesirable in those varieties having a colored crest. When the
crest is plucked, the full strength of the bird will go into
the growth of the new crest feathers, making them longer
than they would be if grown along with the rest of the
plumage.

Injury to the crest is apt to be caused by the females
picking the crest feathers of the male, or of one another.
They are most apt to pick the crest when it is wet or when
the feathers are in the pin feather stage. The injury resulting, if only slight, will cause the feathers to come in
white. It is best to keep the males each by himself except
during the breeding season, for their crests can be com-

pletely spoiled in a very short time. With age, or in other words, with each successive molt, there is a tendency for the crests to come lighter, that is, to show more white.

The bearded varieties of Polish are just the same in every respect as the corresponding varieties of the non-bearded. except in the presence of the beard or muff. The beard should be full, covering the face and wattles almost completely in both sexes. The plain or non-bearded varieties never show any tendency to come bearded, but the bearded varieties show more or less tendency to come with a scanty beard or plain.

In feathering, the Polish is somewhat looser and more profusely feathered than the Leghorn.

In mating this breed, the following defects should be guarded against in so far as possible: too small size; deformed or roach back; white legs; stubs and down; too large and coarse comb; single comb or leaf comb; closed nostrils; lightly feathered or whiskbroom-like crest; split or parted crest; twisted feathers in front of crest falling forward over the bill; crest set crookedly on the head; and scanty beard or absence of beard in the bearded varieties.

The White-Crested Black Polish

This variety, which is non-bearded, is one of the best in type and general excellence. In mating, the single or standard mating only is employed. Birds are selected as breeders which approach as nearly as possible the standard requirements, both in type and color. The black body color of both sexes should be black with a green sheen, and should be free from purple barring. Purple barring may occur in birds as the result of breeding, or may be due to the poor condition of the birds when the feathers were developing, or may be due to exposure to the rain and to the sun. When due to some other cause than the breeding of the birds, it is not nearly so serious a defect. Birds showing purple barring due to poor condition or exposure may often molt in without showing purple barring by keeping them out of the weather.

White may occur in under color of hackle of males and must be guarded against. Another common defect is white in the wings, which may be so bad as to show entire white wing feathers in either sex. The wings and tail of females sometimes show a little gray. If this is due to injury, as sometimes happens, it is not a breeding fault, but if not due to injury, it must be looked out for. The feathers of the female around the vent are sometimes tipped with pale gray, which is, of course, undesirable. It is necessary to guard against too many black feathers in the front of the crest, and breeders should be selected which are free from them, or which have the smallest possible number.

Sometimes males are obtained which show a little brass or straw color in hackle or back. Such males should not be used for breeding unless it becomes necessary to make a mating for the purpose of improving the green sheen of the females. In such a case the males showing a little brass or straw color should be mated to females with as good a sheen as possible, as this will result in an improvement in the females in that particular.

In mating this variety, the following defects, in addition to those common to the breed (page 232), must be guarded against in so far as possible: purple barring; white in under color of hackle of males; white in wings or entire white wing feathers in both sexes; gray in wings and tails of females; feathers about the vent of females tipped with gray; too many black feathers in the front of the crest; brass or straw color in hackle or back of males, except for special matings.

The Bearded Golden Polish

In mating this variety it is usual to employ the single or standard mating. Birds of both sexes are selected as breeders which approach as nearly as possible to the standard. The lacing should be a narrow, lustrous black, which extends clear around the outside edge of the feathers, leaving a large, clear, open center of rich, red bay color. There should be no edging of golden on the outside of the black lacing, as sometimes occurs, which is often referred to as frosting. The lacing should approach as nearly as possible to the character of the Sebright lacing, but is not quite as narrow. It is necessary to look out for spangling of the feathers instead of lacing. It is better to have the feathers show moon-shaped marks rather than spangles, although the former marking is not desired. A peppering of black in the golden color of the feathers, which is often called mossiness, frequently occurs, especially in females, and is undesirable. Black peppering in wings and tail of both sexes must also be guarded against, as must white in the flight and main tail feathers.

In pullets of both the Golden and Silver Polish, the color pattern of the crest feathers is just the opposite of that of the body feathers. Instead of the golden or white centers and black lacing of the body, the crest feathers are black with a golden or white lacing, according to the variety. As the pullets molt in as hens, the color pattern of the crest feathers should reverse so that it corresponds to that of the body feathers. This reversal does not always occur, however, only females of the right blood lines showing the change. See Fig. 79.

In making the mating it is especially important that the male should be a good rich red bay in color, free from frosting, in order to keep up the color of the females. The female should be as near the male in color as possible. A good guide for color is to have the breast of the male harmonize with or match the ground color of the female. If lighter birds are bred, they are apt to throw white flight feathers.

It is necessary to avoid black ends to the lower back feathers and tail coverts. Some females come too dark for exhibition as pullets. However, there is a tendency for them to lighten with age, and they often make the best hens. White in crest must also be avoided. This is apt to increase with age and is in consequence not as bad a breeding fault in old males as in cockerels. Where the white is due to injury it need not be considered from a breeding standpoint. In mating this variety, the following defects, in addition to those common to the breed (page 232), must be avoided in so far as possible: too heavy lacing; frosting; spangling or moon marks instead of lacing; black peppering or mossiness in the ground color; black peppering in wings and tail; white in flight and main tail feathers; general color of plumage too light; black ends to lower back feathers and tail coverts; and white in crest.

The Bearded Silver Polish

In mating this variety, only the standard or single mating is used. The general considerations of the color mating are exactly the same as in the Bearded Golden Variety, except that the golden ground color is replaced by white. It is important not to breed males with bronze or rusty hackle, back, wing bows or saddle, as they will be apt to get females with peppering in the tail and with feathers lacking a good white center and having a frosty edging.

A mating is sometimes made of a Silver Polish cock and a White Polish hen for the purpose of securing the laced feathers with pure white centers. Use a male from this mating free from rustiness to breed back to the Silver females. This is done to get rid of the smutty center.

For defects which must be guarded against, see the defects common to the breed (page 232), and the defects for the Bearded Golden Polish (page 235).

The Bearded White Polish

In mating this variety, the single or standard mating is used, that is, a mating in which the breeders of both sexes are as near standard as possible. As in many other white varieties, there is likely to be some trouble with brassiness

Fig. 78—Well-marked Silver Polish feathers. M indicates male and F female. (Photograph from the Bureau of Animal Industry, United States Department of Agriculture.)

in males, and creaminess in females. Avoid breeding from males which show brassiness. Another defect which sometimes occurs, which is not as troublesome in White Polish as in some other white varieties is black ticking. However,

Fig. 79—Silver Polish pullet. Notice that the markings of the crest feathers are the reverse of the body feathers. As the pullet molts in as a hen, the markings of the crest feathers will change if the breeding is right so as to correspond with that of the body plumage. (Photograph from the Bureau of Animal Industry, United States Department of Agriculture.)

this should be selected against in picking out the breeders. Red feathers occasionally occur in the shoulders and backs of males, and this must, of course, be avoided. There is some tendency toward white legs, and to offset this, use breeders which have strong blue legs.

For defects common to the breed, which must be guarded against, in so far as possible, see page 232.

The Buff Laced Polish

This is a bearded variety. There is also a plain or non-bearded Buff Laced Polish which is not as yet a standard variety. In breeding, a single or standard mating is employed. The general color scheme is the same as the Silver or Golden varieties, except that the ground color is buff and the lacing is pale buff, shading into white. The character of the lacing should approach as nearly as possible, as in the Golden and Silver, to the Sebright lacing, but is not quite as narrow. It is desired to get the pale buff lacing as near white as possible. It is also desired to secure fowls which are the same shade throughout, but it is seldom possible to get birds perfect in this respect. The males tend to run too dark in buff, often showing red on shoulders, back and wing bows and black feathers in wings. Such males, when bred, are apt to get females with light hackles and dark bodies. A male real light in tail is apt to get good lacing in his daughters. There is also a tendency for most females to run a trifle light in ground color. Black feathers and black peppering in wings, tails and tail coverts must be avoided in both sexes. This variety is inclined to run too small in size and to be deficient in size and shape of crest.

In mating this variety, the following defects, in addition to those common to the breed (page 232), must be guarded against in so far as possible: uneven color; red in shoulders and backs of males; black feathers in wings and tail; females a trifle too light in color; black peppering in wings, tail and tail coverts; too small size; too small crest and poorly shaped crest.

The Non-Bearded Golden, Silver and White Polish

These non-bearded varieties are identical in every particular with the corresponding bearded varieties, except for the absence of beard. Exactly the same principles hold in making the matings, except to use breeders which have no beards.

CHAPTER IX

THE HAMBURG CLASS

The Hamburg

This is a breed of rather small size, being, if anything, slightly smaller than the Leghorn. In type, it resembles the Leghorn, having nearly the same shape of body, and shows the hock joint and a part of the thigh distinctly. The birds are very neat, well rounded and well finished throughout. The body is carried nearly level, the back line being practically so in the female, but with a slight slope downward from the shoulders to the tail in the male. The back should be moderate in length. The breast is prominent and well rounded. The tail is carried comparatively low and any tendency toward a high tail must be avoided. The tail feathers are long and well spread, and in the male very profusely furnished with long, curving sickles and with profuse tail coverts. Any tendency toward short sickles should be guarded against, except where hen-feathered males are bred.

The shanks should be free from any stubs or down. Most of the varieties are not much troubled in this respect, but the Golden and Silver Spangled show some tendency toward stubs. The shank and toe color is leaden blue, with black preferred in the Black variety. The feathering of the Hamburg is medium close.

The Hamburg comb is rose and rather large for the size of the bird, but too beefy combs must be avoided. The comb should be a typical rose comb, square in front, set firm and even on the head, and free from hollows either on top or along the sides. The spike should be well developed and should have a slight upward turn at the end, as this is characteristic of the Hamburg. Too broad a comb, one which is lopped or one which has no spike should be avoided in breeding.

The ear lobe is white and nearly round. It should be free from red, but, as in the case of other white ear lobes, red may develop with age, so that its presence in the lobe of a cock or hen which was sound as a cockerel or pullet is not so serious a breeding fault as red, which is present in the lobe of a cockerel or pullet. The ear lobe should be medium to rather large in size, and should be flat and smooth. Too large an ear lobe is undesirable, as it is apt to be associated with white in face. The face should be red and the line between the face and ear lobe should be distinct and clean cut. White in face is a common and serious defect, but, like red in ear lobe, may develop with age, in which case it is not so serious a breeding defect as when present in the young stock.

In Hamburg chicks there is a tendency for the feather growth to outstrip the body growth, particularly in the wing feathers. It is, therefore, necessary to give them stimulating feed in order to keep up the body growth. It is sometimes necessary to clip the wing feathers where this occurs.

In breeding Hamburgs, the following defects must be guarded against in so far as possible: high tail; tail not well spread; tail poorly furnished; short sickles; stubs and down; shank and toe color other than leaden blue or black in the Black variety; too large and beefy comb; hollow comb; spike not turning up at the end; lopped comb; absence of spike on comb; too large an ear lobe; wrinkled ear lobe; red in ear lobe, and white in face.

The Golden Spangled Hamburg

In breeding this variety, it is most common to use a single or standard mating. Select a male which is a trifle richer or darker in ground color than the standard and mate him to standard colored females. From such a mating will be obtained both males and females of good quality. The contrast between the golden ground

Fig. 80—Well-marked Golden Spangled Hamburg feathers. M indicates male and F female. (Photograph from the Bureau of Animal Industry, United States Department of Agriculture.)

color and the black of the spangles and the hackle and
saddle striping must be distinct and the line between
the two colors clean cut. The shape of the spangle is
also important. It should cover the end of the feather
and run up the feather to a point at the quill, that portion
of the spangle being V-shaped with the apex of the V
at the quill. Frequently the spangles may fail to run to
a point and may be either crescent or moon shaped, or be
round or circular in shape. Such spangles are undesirable. It is more difficult to get proper shaped spangles
on the Golden Spangled than on the Silver Spangled variety, especially in females. The black spangles should
show a distinct green sheen or luster. The spangling
should be uniform all over the body, bearing in mind the
difference in the size of the feathers in the different
sections.

Patchiness, that is, black or dark patches in the
plumage, due to the uneven distribution or the variation
in the size of the black spangles, must be avoided. There
sometimes occurs a frosting or edging of golden on the
outside of the black spangle. This is very undesirable.
Shafting may occur in both sexes and should be selected
against. It must be kept in mind that the exhibition
male of this variety has a solid black tail, instead of a
golden tail spangled with black, as might be expected.
In this respect it is different from the Silver Spangled
male, which has a white tail spangled with black. The
hackle feathers in both sexes of the Golden Spangled variety are striped with black, while in the Silver Spangled
variety the feathers of this section are marked with an
elongated black spangle. See Figs. 80 and 81. The
under color should be slaty black. Males having this
under color with a light under color of fluff produce the
best cockerels.

In mating this variety, the following defects in addition to those common to the breed (page 240) must be

guarded against in so far as possible: round, moon-shaped or poorly-shaped spangles; patchiness; frosting; shafting; laced breast in males.

Fig. 81—Well-marked Silver Spangled Hamburg feathers. M indicates male and F female. (Photograph from the Bureau of Animal Industry, United States Department of Agriculture.)

The Silver Spangled Hamburg

In breeding this variety, exactly the same method is used as in breeding the Golden Spangled variety. These two varieties are identical except that the Golden color of the Golden variety is replaced by white, and the male tail of the Silver Spangled variety is white spangled with black, whereas that of the Golden Spangled male is solid black; also in this variety the hackle feathers of both sexes are marked with an elongated spangle instead of being striped as in the Golden Spangled variety. The desired shape of spangles is somewhat easier to get in this variety than in the Golden Spangled. There are certain defects which must be guarded against in this variety which do not apply to the Golden Spangled variety. Brown on the throat and shoulders of males and on the throat of females must be avoided. The necks in both sexes are likely to come too light, or even white, and this same trouble is likely to extend around on the throat,

Fig. 82—Hen feathered Silver Spangled Hamburg male. (Photograph from the Bureau of Animal Industry, United States Department of Agriculture.)

In the Silver Spangled Hamburg there sometimes occur hen feathered males. These males are marked like the females, and have short or no sickle feathers and no saddle feathers. Often they will not fertilize eggs, but when they will they should be bred to standard females, as they are especially valuable to produce good females. See Fig. 82.

Where white in face is troublesome, gypsy or dark purple-faced females may be used in the mating to get rid of this defect.

In mating this variety, the following defects in addition to those common to the breed (page 240) must be guarded against in so far as possible: broad, moon-shaped or poorly-shaped spangles; patchiness; brown on throat of both sexes; brown on shoulders of males; frosting; shafting; too light colored or white necks or throats in both sexes; laced breast in males.

*The Golden Penciled Hamburg

In this variety the penciling should be narrow and straight across the feathers, distinct and clean cut, and the black and bay bars equal in width. The black of the penciling should show a green luster or sheen. The hardest places to get good penciling in the females are on the throat and breast. Patchiness, or an un-

Fig. 83—Cockerel bred Golden Penciled Hamburg female showing coarse, irregular penciling. (Photograph from the Bureau of Animal Industry, United States Department of Agriculture.)

Fig. 84—The markings of a hen feathered male are nearly identical with those of the females of the same variety. In the case of the male from which these feathers were taken, all sections are hen feathered except the hackle, the feathers of which are the same, both in shape and markings, as those of males which are not hen feathered and which are standard in hackle. Males occur in which the hackle also is hen feathered. See Fig. 87. (Potograph from the Bureau of Animal Industry, United States Department of Agriculture.)

evenness of the general ground color, occurs, but must be avoided. In the standard male there is some tendency for the black and golden in the sickles and coverts to mix. These feathers should show a sharp, clear-cut banding or edging about ⅛ inch wide, with a rich, lustrous, clean black center.

In breeding this variety, it is customary to use the double mating system. The two matings are as follows:

Cockerel mating.—Select a standard male. To him mate females whose penciling is very coarse and irregular. Such females usually are lacking in penciling of breast and throat. See Fig. 83. Be sure that their flight feathers are a good, positive black, as this gives the black ground color and strength of color to the male's tail.

Fig. 85.—Pullet bred Golden Penciled Hamburg female showing fineness and regularity of barring as compared with the cockerel bred female, Fig. 83. (Photograph from the Bureau of Animal Industry, United States Department of Agriculture.)

Pullet mating.—Use a hen-feathered male if possible, that is, a male which is as near the exhibition female in markings as possible. Males almost identical in markings to the females are produced and are valuable as breeders if they will fertilize eggs. Often, however, these hen-feathered males will not fertilize. These males have short sickles or no sickles, which increases their resemblance to the females. Use standard colored females for the mating. See Fig. 85.

In breeding this variety, the following defects in addition to those common to the breed (page 240) must be guarded against in so far as possible: Too open penciling, penciling not straight across the feathers, patchiness or unevenness of color, the black and golden of the sickles and tail coverts mixing, and poor penciling on the breast and throat of females.

The Silver Penciled Hamburg

This variety is practically identical with the Golden Penciled variety, except that the golden color is replaced by white. In breeding, the double mating system is used and the matings employed are the same as for the Golden Penciled varieties (page 247), bearing in mind the difference in color of the two varieties.

---

*Confusion sometimes arises over the term penciled or penciling, due to the fact that it is used to describe two different types of markings. In the Silver and Golden Penciled Hamburgs, the penciling consists of a series of parallel bars extending across the feathers, while in other penciled breeds, such as the Silver Penciled Wyandotte, the Silver Penciled Plymouth Rock or the Dark Brahma, the penciling consists of a series of concentric parallel markings, which follow the shape of the outline of the feathers. This latter form of penciling is like that which is found in Partridge varieties. (See page 245.)

Fig. 86—Well-marked Silver Penciled Hamburg feathers. M indicates male and F female. (Photograph from the Bureau of Animal Industry, United States Department of Agriculture.)

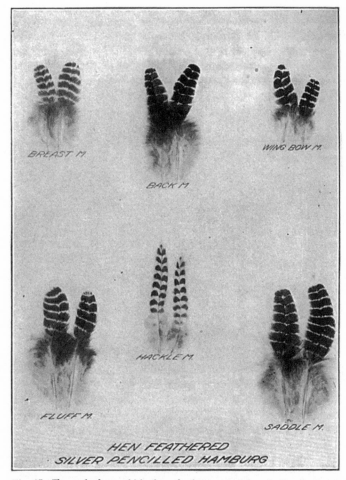

Fig. 87—The male from which these feathers were taken is hen feathered in all sections, as is evident both in the shape of the feathers and the markings, which are like those of Silver Penciled Hamburg females. (Photograph from the Bureau of Animal Industry, United States Department of Agriculture.)

The White Hamburg

In breeding this variety, the single or standard mating is employed. Birds of both sexes are chosen as breeders which most nearly approach the standard. As in any other white variety, the plumage should be as pure white as possible, free from any brassiness or creaminess, and from any foreign color. Black ticking and brassiness are the two most troublesome color defects in this variety. Guard against pale or white legs in both sexes. For defects common to the breed which must be guarded against in so far as possible, see page 240.

The Black Hamburg

This variety is especially characterized by the green sheen of the birds of both sexes, which is so pronounced as to cause good specimens to appear almost beetle green in color. This sheen should be present in the females as well as the males, and the Black Hamburg females are probably ahead of any other black females in the height of green sheen and the freedom from purple. Purple barring is often largely a matter of condition, and birds in perfect health seldom show it. However, purple barring not due to poor condition is present in some strains and is, of course, a defect.

In breeding this variety, both the single mating and the double mating systems are employed.

In the single mating, use a highly colored beetle green male with black females of medium sheen. From this mating will be secured good specimens of both sexes. This mating is recommended because it will produce the fewest culls.

Where double matings are used, the pullet mating should consist of a high-colored male which may show a little red or straw color in hackle and saddle, mated to high-colored females. From this mating the females will be very fine, but the males will often come with red or straw colored hackles. These males can be used in the future pullet matings. To get extreme sheen in females, a Golden Spangled male is sometimes mated to high-colored females. The males from this mating are, however, useless either for exhibition or breeding.

The cockerel mating should consist of an exhibition male mated to real dull-colored females. This mating will produce a fair percentage of fine colored males and also a smaller percentage of good females. The females for this mating usually have a longer tail, with long and slightly curved sickle feathers. Where white in face is troublesome, gypsy-faced females can be used in the cockerel mating to get rid of the defect. For defects common to the breed which must be guarded against in so far as possible, see page 240.

CHAPTER X

THE FRENCH CLASS

The Houdan

This breed is comparatively low set in type, approaching somewhat the Dorking in this respect. The body is long and broad, and the breast full and prominent. This breed has a large, full crest, and possesses a beard. The comb should be V-shaped and small. Any tendency toward a large comb must be avoided. The comb should be set well back against the crest in both male and female. The smaller the comb the better. The beak should have an open or raised nostril (page 253). Fig. 88.

The tail is moderately low carried and is much fuller and better spread in the male than in the female. The body slopes slightly from the shoulders toward the tail. It is important that the crest be set firmly and straight on the head, with no tendency to be lop-sided. Especially large crests

Fig. 88—Head of Houdan male. 1—Crest. 2—V-shaped comb. 3—Raised or open nostril. 4—Muff. 5—Beard.

are very apt to be lop-sided, or, in other words, to fall over to one side.

The fourth and fifth toes should be well separated, show-ing no tendency for these two toes to grow together. The legs and toes should be free from feathers, but there is some tendency toward stubs. Frequently specimens show down between the toes, and this is apt to be more apparent as the age increases. It is not necessary to discard for breeding purposes birds which show down, although, of course, it is better to use birds which are free from this.

In making the matings in this breed it is necessary to guard against the following defects in so far as possible: body set too high on legs; too large comb; lop-sided crest; fourth and fifth toes growing together; stubs.

The White Houdan

In selecting the matings for this variety, a single or standard mating system is used. The plumage of both sexes should be white throughout and should be free from foreign color of any kind. For defects common to the breed, which must be guarded against in so far as possible, see page 254.

The Mottled Houdan

In making the mating for this variety it is common to use the single or standard mating. Birds of both sexes should be selected which approach the standard as closely as pos-sible, both in type and in color. There is, however, many times a tendency for the hens to become lighter in color as they grow older, this being due, as in the case of other mottled varieties, to the appearance of more white with successive molts. If this tendency for the hens to come too light is troublesome, it must be offset by using darker males to mate with the hens. The white tip which forms the mottling should be V-shaped, and the white and the black should be distinct, there being no tendency for the black to run into the white. Any tendency toward round tips instead of V-shaped tips must be guarded against. There is also a tendency for the tips to come irregular in their shape and to be so large as to cause a white or grayish white splashing through the plumage. This is undesirable, and birds which show much splashing should not be bred from.

White feathers are apt to come in the crest, and the tendency is for the crest to grow whiter with age. Too much white in crest or entire white feathers, especially in young birds, must be selected against. Solid white feathers sometimes occur in the tail and wing flights. It is best not to use these birds as breeders, but if one has exceptional specimens with some white in wings and tail, it is possible to use them by mating with a bird very strong in this respect to offset the defect. White feathers also occur at times in the body of females, and these are undesirable. For defects common to the breed, which must be guarded against in so far as possible, see page 254.

The Crevecoeur

This is a black breed which is slightly larger in size than the Houdan, but which is somewhat similar in type, running a little higher on legs. In crest and comb it is like the Houdan. Unlike the Houdan, however, this breed has but four toes.

There is somewhat of a tendency for this breed to come under weight, and this must be guarded against in making the mating. It is common to use only the single or standard mating, selecting birds of both sexes which approach the standard as nearly as possible.

The commonest color defects are white in wings and white in crest. White occurs in crest with age and rarely appears in cockerels or pullets. It is therefore a much more serious defect when found in young birds than in old birds. White in body plumage should be rigidly selected against, either in young birds or in old birds.

The La Fleche

The La Fleche is the largest of the French breeds which are standard in this country. In type it is a higher, rangier bird than either the Houdan or the Crevecoeur, having a good length of leg. The back is long and broad and the breast full and prominent. The body slants backward from shoulders to tail. The tail is large and well spread, and is carried moderately low in the male and somewhat higher in the female. It is rather difficult to keep this breed up in size, and it is therefore necessary to use birds which are of good size for breeding. The comb should be a clean, V-shaped comb of medium size. It should be larger at the base and gradually taper to two distinct points. There is a tendency, however, for the comb to be beefy and the prongs to flatten out at the ends, thus approaching a leaf comb, which should be avoided. It also shows some tendency to lop forward instead of being erect. This breed should be as free as possible from any sign of crest, which may occur as a tuft of feathers at the rear of the comb. However, it is almost impossible to secure birds which are absolutely free from any indication of a tuft. White in face sometimes occurs, especially in cockerels, and must be avoided.

Breed from big-boned and rangy birds which have long backs and show no tendency toward narrow backs. The beak should be good and strong, with open or raised nostril (page 253). A good width of feather is desired. In making the mating, select birds which approach in both sexes as closely as possible to the standard, both in type and in color.

As in most black breeds, there is a tendency for the plumage to show purple barring. This must be avoided, as the La Fleche is not so free from this as some of the other black breeds, such as the Hamburg and Langshan. White also may occur in under color of hackle, back and saddle, and this is undesirable. Red or straw sometimes occurs in the hackles of males. This should be avoided, unless it is desired to use such males in a special mating for the purpose of increasing the green sheen of the females, which has a tendency to be lacking, and the plumage to become dull black.

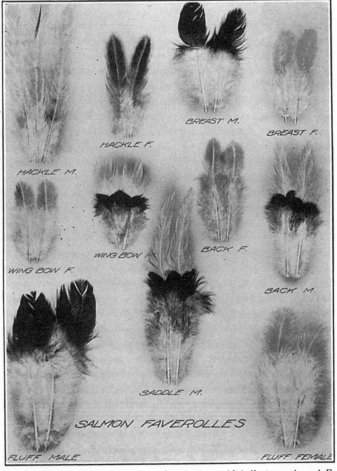

Fig. 89—Well-marked Salmon Faverolles feathers. M indicates male and F female. (Photograph from the Bureau of Animal Industry, United States Department of Agriculture.)

In such a case, use a male with red in hackle with females which show a good green sheen. Guard against white in any part of the plumage, especially in the wing.

This breed is an old established one, and breeds comparatively true to standard requirements.

The Salmon Faverolles

The Faverolles is a breed of which there are several varieties, but at the present time the salmon variety is the only one which has been admitted to the Standard of Perfection. This breed is broad and deep in body, but has not the length of body which the other French breeds possess. It is free from crest, but has a full beard and muffs. The Faverolles also has a fifth toe, and the shanks and the outer toes may be sparsely feathered.

In making the mating it is usual to employ the single or standard mating. One of the principal color defects which must be guarded against is striping in the hackle and saddle of the male. Males without any stripe in hackle and saddle give a clear hackle in females, while those which show striping are apt to cause a black ticking in the hackle feathers. Breeders of both sexes should have clean hackle. The breeders should be set quite low on legs, and it is necessary to guard against fowls which are too long in leg. Birds which are of too upstanding a type and which lack in breast and length of keel, and those showing only four toes, must be selected against. Vulture hocks sometimes occur in males, but this defect is not so troublesome in the Salmon Faverolles as in the other non-standard varieties.

CHAPTER XI

THE CONTINENTAL CLASS

The Campine

This is a breed which in size and type approaches very closely to the Leghorn. It possesses a long, rather slim body, tail carried moderately low, and set quite high on legs. The comb is single and is like the Leghorn comb except that it has somewhat more of a tendency to run too large. The male of this breed differs from the Leghorn male somewhat in the character of feathering. Instead of having long, flowing saddle feathers like the Leghorn male, the Campine male has a hen-feathered saddle, in which the feathers are relatively wide and have round tips instead of pointed tips.

In making the mating in this breed, it is necessary to guard against high tails, which are quite troublesome. It is also necessary to be sure that the eyes are dark brown, or if possible, approaching black in color, and do not tend toward a lighter red eye, as there is a decided tendency in older birds for the eye to fade out to a lighter color. As in the Leghorn, there is some trouble with white in face, particularly in the male, and this must be guarded against. Stubs sometimes occur, but they are not very troublesome in this breed. The back should be long, and this, together with the medium, low-carried tail, does away with any appearance of a "U"-shaped back. Birds with such a back are not desirable as breeders.

Occasional tinted eggs will be obtained from females of this breed. As chalk-white eggs are desired, hens should be selected as breeders which lay eggs of this color, and the male should be out of a hen which lays chalk-white eggs.

Fig. 90—Well-marked Silver Campine feathers. M indicates male and F female. (Photograph from the Bureau of Animal Industry, United States Department of Agriculture.)

The Silver Campine

In selecting a mating for this variety, use a single or standard mating. Birds of both sexes should approach the standard as closely as possible, and the barring of the feathers of the male and female should be as nearly alike as possible. It is not good policy to mate a bird of one sex having a narrow barring with birds of the opposite sex having a wide barring. The barring should be sharp and as clearly defined as it is possible to secure it, and the colors should not mix or run down the quill. While the black bar should be slightly V-shaped with the point of the V at the quill, this V-shape should not be very marked. Females which show nice, clean breast barring are quite apt to have a slight amount of ticking in the lower half of the hackle. It is best not to discard pullets showing this ticking, but otherwise good, as they are apt to molt in with clear hackles, as the hackle tends to get whiter at each molt, while the breast barring is likely to hold good. This ticking occasionally occurs in males also, and this should be avoided so far as possible. There is a tendency for white saddle hangers to occur in the male, and this must be avoided. Extremely wide back feathers when they occur indicate the Belgian type, and are not desirable. As stated before, the barring should be distinct and well defined, and there should be no indication of an intermediate brown between the black and white bars. It is difficult in both sexes to get the main tail feathers barred clear across, although the sickles of the male come good in this respect.

For defects common to the breed which must be guarded against in so far as possible, see page 259.

The Golden Campine

This bird is exactly the same as the Silver Campine except that the silver color is replaced by golden bay.

In making the mating, the same considerations should be kept in mind as in the Silver Campine (page 261), always remembering this fact—that the silver is replaced by the golden bay color. For defects common to the breed which must be guarded against in so far as possible, see page 259.

CHAPTER XII

THE GAME AND GAME BANTAM CLASS

The Exhibition Game

Exhibition games are to the chicken world what the thoroughbred is to the horse world. They must be full of stamina, vigor and health. They must be racy, keen of eye, alert, active and high strung. Their general poise and spring may be well expressed by saying that they must be on tiptoe all the time. Weak constitutions must be guarded against, as such birds will not have the alertness and snappiness desired.

The high station of these birds is very characteristic of the breed. Their bodies are set high up on long legs which show length, particularly in the thigh. The shank and by far the greater part of the thigh is distinctly visible underneath the body. These birds have by far the highest station of any of the standard breeds. The maintenance of this high station, since it is so characteristic, is very important.

The head should be long and slender, and any tendency toward a short and thick head, or, as sometimes termed, a "bull-head," is to be avoided. There should be no prominence or projection of the skull over the eyes. The comb of the male is unimportant, since the males are dubbed at an early age. However, a small, erect and thin comb, with even serrations, is desired on the females. The eye must be fiery and show lots of spirit. Avoid a mellow eye, that is, one lacking in spirit and having too much of a soft or mild quality.

The neck in both sexes is long, slender and erect, and blends smoothly with the back. The hackle feathering is close and rather scanty, especially in front.

In general, the body is well rounded. There are no conspicuous angles in the make-up of the bird. Viewed from the side, the body has the general shape of a beef's heart, with the large end at the shoulders and the small end at the stern. The body and the back line, which is straight, slopes downward decidedly from front to back. The back is broad at the shoulders and tapers back toward the stern, giving the body a flatiron shape as viewed from above. Avoid a bird as a breeder which does not taper in body, as from that type one cannot expect to produce good specimens. The saddle is short, the ends of the saddle feathers not extending below the underline of the body. The breast is quite prominent, well rounded and carried well up. Crooked breasts must be avoided.

The wing should be short and well tucked up, that is, held up snugly in place. The wing should lack about half an inch of reaching the stern. The Black Red and Red Pyle varieties are best in this respect, while the Black and the White varieties are the poorest. Wings are sometimes carried with the points up on the back. This should be avoided.

The stern must be well tucked up and well muscled. It must not be loose or have a tendency to hang down. Of course hens with age and when laying show more of a tendency toward this condition.

The tail must be closely and tightly folded and run out to a point, or be what is often termed a "whip" tail. Do not breed from loose-tailed birds, as the shape of tail is very important. The tail is carried very low, just about on the horizontal. High tails are undesirable. The junction of the back and tail is smooth, no angle being caused.

The legs must be long, straight and wide apart. The birds must be absolutely plumb or square on their shanks or "pins." There is some tendency not to be set square on legs, and this must be avoided in the breeders. The legs must be set wide apart, with no tendency toward knock-knees. The shanks should be round, like pipe stems. Avoid any tendency toward flat shanks. The toes must be firm and straight and be of good length, but not so long as to be out of proportion, as the bird must have grace. It is important that the fourth toe extend back, as this helps to give the bird poise and spring. Avoid a fourth toe which bends around toward the side, or the condition which is commonly known as "duck foot." The birds must stand strongly on legs and the hock joints must not be weak or show a tendency to buckle. Many specimens otherwise nearly perfect have such weak legs or hock joints that they cannot get up.

The bone of this breed should be good, hard and straight in every section, but while a good quality of bone is necessary, it should not be coarse. The birds should also be hard and well muscled throughout.

The texture of the feathers and the character of the feathering are very important. The feathers should have hardness and shortness, and the quills must be hard. This is very important, as the feathers must hug the body closely so as to give the extremely tight feathering desired. There must be no looseness of feathering. When the feathers are raised in any section and then released, they should snap back into position like springs. The feathers of the wing, when riffled, must show hardness and snap back into position sharply.

The birds for breeders should run good in size for the breed. Too small birds should be avoided, and very large birds are not apt to be as good in type.

Great care must be exercised in raising and handling games in order to get the best results. This is but little less important than the breeding, in so far as type is concerned. They must be carefully handed until they are through the pin feather stage, as they are delicate up to this point. For the same reason they must be kept free from lice, as they cannot stand them. Until they have gotten a good start, they must be kept out of the morning dampness and dew. To keep them high strung and to avoid crowding, which they cannot withstand, they should be raised only in small flocks. They should be raised on range to give them constitution and good bone. As they grow, they should be fed in cups, placed high on the wall so as to give the birds reach. When being trained also, this high feeding should be continued. Hard grain only should be fed, never any sloppy feeds, so as to develop bone and muscle and to make the feathering tight and hard. Overfeeding must be avoided, as it tends to cause the birds to go weak on legs. They must not be allowed to roost very early, and when roosts are provided they should be broad, in order to avoid crooked breast bones. The birds must be held carefully and securely when handling them, as they are likely to flutter or struggle suddenly and violently, and with a quick twist may break their legs.

When exhibiting, it is well to keep feed from the birds until after they are judged, as feeding tends to destroy the lines of breast by causing the crop to protrude.

When the birds are in condition to show, they should have a hard-muscled feeling, or, as often stated, should be hard as nails. There should be an entire absence of any flabby feeling.

In breeding Games, the following defects must be guarded against in so far as possible: lack of constitution, vigor, stamina or health; lack of activity and alertness; listless poise; lack of high station; too short legs; short, thick head; prominence or projection of skull over eyes; mellow eye, that is, one lacking in fire and spirit; body or back not tapering from shoulders to stern; crooked breasts; wing too long; wing not well tucked up; loose tail, that is, one not closely and tightly folded; birds not square on legs; legs not straight; knock-knees; flat shanks; duck foot; weak or buckled hock joints; loose or baggy stern; too coarse bone; not well muscled; feathers not hard and short; feathering not tight; too small birds; extremely large birds; too high tail and wings carried up on back.

The Black-Breasted Red Game

In breeding this variety it is customary to use double matings. However, there is some difference in the matings used, and two distinct sets of matings are here described.

One breeder advises the following mating:

Cockerel mating.—Select a standard-colored male. Guard against pronounced hackle and saddle striping in exhibition Black Red males. With him mate females which show excessive steely gray stippling on back and some red, bricky or wheaten color on wing bows and shoulders. These females should be standard in other respects.

Another breeder advises a cockerel mating in which both the male and the females are standard. The percentage of standard-colored males from this mating is much greater than of standard-colored females.

Pullet mating.—The pullet mating advised by the breeder giving the first cockerel mating above, consists of a male which is generally darker in top color, that is, hackle, back, wing bows and saddle, but free from red in wing bows. He should also show some red coloring on breast and fluff, this being more pronounced as the bird ages. This red in breast and fluff is a distinct difference from the cockerel-bred male, which should be solid in his black at all times. The male for this mating should also be as free from striping in hackle and saddle as possible, although a small amount of striping in these sections is not serious. The females used in this mating should be standard.

The pullet mating advised by the breeder giving the second cockerel mating above consists of a male showing a very light top color throughout, one which approaches lemon very closely and gets away from the red. He should be sound, that is, solid black, in breast and body color. Mate standard-colored females with this male. This mating will tend to keep down the stippling in the females so that it will not be too strong and will produce a larger percentage of the grayish brown females which are so desired.

For defects common to the breed, which must be guarded against in so far as possible, see page 266.

The Brown Red Game

In breeding this variety, both the single or standard mating and the double mating systems are used. In the single or standard mating, birds of both sexes are selected as breeders which approach the standard as nearly as possible. One of the hardest things in this variety is to keep a dark purple or "gypsy"-colored face. Therefore, it is important that the breeders approach this color of face as nearly as possible and be well away from the red face. A common weakness of males is to run to a dark orange top color. The top color of males should be a deep lemon, gradually shading to a lighter lemon as it runs down on the back and saddle, as this produces the proper-colored lacing of the breast feathers in both sexes. There is a tendency for the back of males and females to be laced. This section should be solid lemon in males and black in females. There is also some tendency for the lacing on breast in both sexes to be too heavy and to extend down the thighs to the shanks. This is undesirable, as a neat, distinct lacing, not too heavy, is desired, and the thighs should be black. The lemon color should run clear over the top of the head in both sexes, as there is a tendency for the top of heads to be black. Shafting in the body plumage, and in particular in the laced feathers of the breast, is not uncommon. This is undesirable, as both the shaft and the centers of these feathers should be black and distinct from the lemon lacing.

Where the double mating system is employed, the matings are as follows:

Cockerel mating.—Use a standard-colored male, showing the rich lemon top color. To him mate females which show a tendency to lemon lacing over the shoulders and back. Such a female is, of course, not standard, as a show specimen should have lacing only on the breast, the back, body and stern being black.

Pullet mating.—To secure a good percentage of standard-colored females, the standard mating just described is used where both the male and the females in the mating are as near standard as possible. While some standard males will also be obtained from this mating, a larger percentage will be secured from the cockerel mating described.

In breeding the Brown Reds, a small percentage in both sexes come white, with blue legs. These are generally regarded as sports and as a rule are discarded as breeders.

In mating, guard against the following defects, in so far as possible, in adition to the defects common to the breed (see page 266): lacing on back of male; lacing on back of female; too heavy lacing on breast of both sexes, extending down the thigh to shank; black caps on heads in both sexes; shafting; red face; and too dark a top color in males, approaching dark orange.

The Golden Duckwing Game

In this variety a good rich golden color is desired on wing bows, hackle, back and saddle of males. This color is difficult to maintain and the principal problem in breeding lies in keeping up the golden color. The usual method of mating is to use a standard male and standard females. Be sure that the male is free from black striping in hackle and saddle and that the breast is free from white ticking and the fluff from frosting. This ticking is apt to appear or increase with age, and is more serious from a breeding standpoint in young than in old males. Also avoid any rusty or bricky color on shoulders and wings of females. This brick color is often called rustiness,

From this mating will be secured standard birds of both sexes, but there will also occur some Silver Duckwing birds. As this line of breeding is continued, the golden color becomes weaker and more and more Silvers will be produced. In order to strengthen the golden color again it becomes necessary to use some Black Red Game blood. Select a standard Black Red male, with rather light top color and with willow legs. Mate him to a very light-colored Golden Duckwing hen or to a Silver Duckwing hen. From this mating will come good Golden Duckwing specimens of both sexes, and these can be used to continue the breeding. As the color begins to fade again, repeat the infusion of Black Red blood.

In mating this variety, guard in so far as possible against black striping in hackle and saddle of males; white ticking in breast of male; frosting in fluff of male; and brick color in wing of females. For defects common to the breed, which must also be guarded against, see page 266.

The Silver Duckwing Game

Birds of the color of this variety come from the Golden Duckwing matings. The Silver Duckwing color will breed true, a standard or single mating being employed. In making this mating, use birds of both sexes which are as near the standard as possible, and avoid any black striping in the hackle or saddle of the male, any white ticking in his breast or frosting in his fluff, and also any brick color on wings or, as sometimes termed, rose wing of females. Care must also be exercised to see that the white is white and not yellow, as there is a tendency in this direction. For defects common to the breed, which must be guarded against in so far as possible, see page 266.

The Birchen Game

The Birchen Game is practically identical with the Brown Red variety, except that the lemon of the latter is replaced

by white. It is necessary to be sure that the breeders
selected show a pure silvery white in lacing and other white
sections, and that it does not run to a yellow tinge. Otherwise the matings and color considerations are the same as
in the Brown Red variety (see page 268), keeping in mind
always that the lemon is replaced by white. For defects
common to the breed, which must be guarded against in so
far as possible, see page 266.

The Red Pyle Game

The great difficulty in breeding this variety is to keep up
the strength of color, particularly the breast color of females and the color of the secondaries or wing bay of males.
There is a tendency for the color to run to a light lemon,
or even to a white breast in females. Such birds of light
coloring are often called Lemon Pyles or White Pyles.

In breeding, a double mating system is employed, together with a special mating from time to time as the color
tends to run out. The regular double matings are as follows:

Cockerel mating.—Use a standard-colored male, especially strong in coloring of the secondaries or the wing bay.
The females should have a solid salmon breast and should
show some red color on shoulders or sides of wings, that is,
be rose-winged. It is no disadvantage if the females show
some red lacing down through the back.

Pullet mating.—Use a male which is a trifle lighter in
color than the exhibition male. He should have a nearly
white hackle, slightly striped with red. Use females of
exhibition color.

In order to strengthen the color, which tends to become
too light, some Black Red blood is introduced from time to
time. When the Red Pyle matings show a tendency to weakness of color in the wing ends of males, it is an indication
that it is time to introduce the Black Red blood again to keep
up the Red Pyle top color to the wing ends. As to how this
may best be done, there is some difference of opinion among

breeders. One breeder advises as follows: Use a standard Black Red male, which must excel in the soundness of sheeny black in the black sections, as the blacker the bird

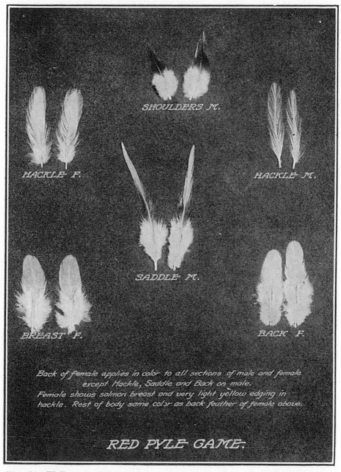

Fig. 91—Well-marked Red Pyle Game feathers. M indicates male and F female. (Photograph from the Bureau of Animal Industry, United States Department of Agriculture.)

the whiter his offspring. He should have yellow legs if possible, but it is hard to find one with yellow legs which is good enough in top color. Mate him to a light breast colored female, that is, a Lemon Pyle female. The pullets from this mating, with a very few exceptions, come black red in color, with yellow legs. Any of these female offspring of exceptional merit are bred to a Red Pyle male of standard color, which has come from a line of Pyles bred straight for several generations. From this mating will be secured good Red Pyles, both male and female, some of which will have willow legs. These can be used to carry on the Red Pyle matings until the color again begins to weaken, when more Black Red blood must be introduced.

Another breeder advises the following procedure in injecting Black Red blood. Select a willow-legged, very light top-colored Black Red male. Mate him with rather weak-colored Red Pyle females, approaching the White Pyles, but with extra strong, rich, dark yellow leg color. The reason for the use of such strong yellow legs in the White Pyle females is that the Black Red blood of the male tends to cause the loss of the yellow leg color, since the standard Black Red is willow-legged, and will throw a good many offspring with willow legs. The offspring of both sexes from this mating tend to be too strong in color, but many males are of excellent quality and show the characteristics of a good Red Pyle show bird. The males which are the cleanest in the white, and as free from black as possible, should be bred to standard-colored Red Pyle females, and this mating will give both sexes of standard color until the color begins to run out again. However, it will take several years' breeding to get Red Pyle color of the best, as the Black Red blood tends to throw black feathers or feathers speckled with black.

For defects common to the breed, which must be guarded against in so far as possible, the reader is referred to page 266.

The White Game

This variety tends to be inferior in type to the other games, except the Black variety. They often lack in length of head and length of leg and also lack in wedge shape of body. They may also lack in shortness, hardness, and closeness of feathers, and often have far too much hackle and tail. They are often too loosely put up. In breeding, the single or standard mating only is used, birds of both sexes being selected as near standard as possible. It must be kept in mind that the legs should be yellow, so that both white and blue legs must be selected against. The color of plumage should be pure white throughout, and any tendency toward brassiness or creaminess, black ticking or any foreign color must be guarded against. For defects common to the breed, which must be guarded against in so far as possible, see page 266.

The Black Game

This variety, like the White, tends to be inferior in type and character of feathering. It frequently lacks in length of head, length of leg especially, wedge shape of body and shortness, hardness and closeness of feathers. It frequently has far too much hackle and tail and is too loosely put up.

In breeding this variety, the single or standard mating is employed, birds of both sexes being selected which are as near standard as possible. In general, the same color consideration obtains here as in any other black variety. The color should be black throughout, free from any foreign color and from purple barring. There is a tendency toward white ticking or lacing in hackle, which must be guarded against. The under color should be a dull black. Females should have a greenish sheen to the surface color, the same as the males. For defects common to the breed, which must be guarded against in so far as possible, see page 266.

The Game Bantam

Both in color and in type, the different varieties of the Game Bantam should be identical with the corresponding varieties of the large Game. The same general principles must be observed in the matings, except as indicated. Of course there is always a tendency for these fowls to come too large, and small birds or those not over standard weights must be selected as breeders. The care and feeding are the same, except that one must be careful not to overfeed or to feed too liberally on bone-forming material in order to keep down the size. Some breeders keep the birds on about half rations for the same purpose. Breeding must be depended upon mainly, however, to keep down the size. While some breeders hatch late to reduce size, others claim that it is better to hatch at the normal time, not later than June, as otherwise the proper reach and the proper feathering will not be secured. There is a tendency for the feathers of the tail, and for the sickle feathers especially, to come too wide. These are often one-half inch wide, but should be narrower if possible. For a detailed description of the type desired, together with the defects which must be guarded against in so far as possible, the reader is referred to the material on the Exhibition Game (see page 266).

The Black-Breasted Red Game Bantam

It is customary to double mate in this variety.

Cockerel mating.—Use a standard male. Mate him to females from a cockerel mating or line which runs red or bricky over the shoulders and wing bows. Occasionally a wheaten-colored hen, mated to an exhibition-colored Black Red male, will get fine exhibition-colored Black Red males. Great care should be used not to introduce any of these pullets into the pullet line, as they will ruin the color.

Pullet mating.—Use a male darker in color than the standard. A small amount of striping in hackle and saddle is not objectionable in this male. The females used should be standard.

In this variety, particularly guard against too wide sickle feathers and bent hocks.

The reader is further referred to the material on mating the large Black-Breasted Red Game (see page 267), and to the material on the Game Bantam (see page 275).

The Brown Red Game Bantam

In breeding this variety, use the single or double mating and proceed the same as in the large corresponding variety (see page 268). Guard in so far as possible against: too much lacing, especially on the back of females; black head or cap color in females; too long wings; too long sickle feathers; tail spread too much; too wide tail feathers; and too large and coarse birds.

The Golden Duckwing Game Bantam

The color of this variety is the same as that of the large Golden Duckwing, and as in that variety a standard mating is used until, by the occurrence of an unusual number of Silver Duckwing-colored specimens, it is evident that the color is getting weak. It is then time to introduce some Back Red blood. This can be done by breeding a Silver Duckwing male from the Golden Duckwing matings to a Black Red hen. This will produce good cockerels, which can be bred into the Golden Duckwing matings and will serve to strengthen the color. Further information on breeding for Golden Duckwing color will be found on page 269.

In breeding this variety, guard against the following, in so far as possible: black striping in hackle of male; rusty color across lower part of shoulders and bricky color in wings of females; too wide tails; too wide tail feathers; ticking of white in breast of males, which is more prevalent with age,

The Silver Duckwing Game Bantam

This variety corresponds exactly with the large Silver Duckwing and is bred from a single mating in the same way (see page 270). The Silver Duckwing Bantams breed true and better than do the Golden, but the same general defects must be guarded against, in so far as possible, as in the Golden Duckwing Game Bantams (see page 276).

The Birchen Game Bantam

As in the larger varieties, the Birchen is the same as the Brown Red, except that the lemon is replaced by white. In breeding, the single or standard mating is employed, and exactly the same things must be taken into consideration as in the Brown Red variety (see page 268). The same things must also be guarded against as in the Brown Red Bantam (see page 276).

The Red Pyle Game Bantam

This variety, like the large Red Pyle, tends to run weak in color after being bred for a while, that is, the female will come white in breast and the males will lose their color in the wing bay. To keep up and renew the color, breed in a yellow-legged Black Red male. The breeding is usually carried on by double mating. The regular mating and the special mating to introduce the Black Red blood is identical with those of the large variety (see page 271). Sometimes, however, a yellow-legged Black Red female is bred to a light top-colored Red Pyle male for the reason that it is easier to get females than males with yellow legs in the Black Red Bantams. From this mating often come willow-legged Red Pyle females. This is just the opposite of the mating used in the large Red Pyle Game, and is used simply as a matter of convenience.

In breeding this variety, guard against the following defects in so far as possible: weak-colored wing bay in males; too light breast color in females; buckled hocks; too heavy-colored females, unless these are used in a cockerel mating, where they will produce good males.

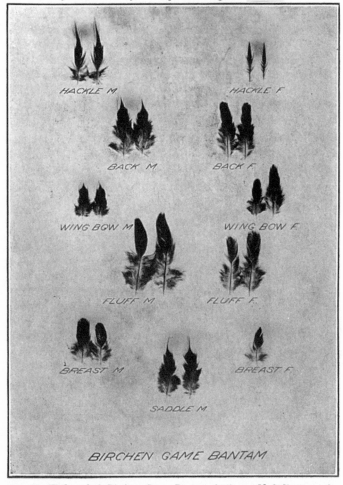

Fig. 92—Well-marked Birchen Game Bantam feathers. M indicates male and F female. (Photograph from the Bureau of Animal Industry, United States Department of Agriculture.)

The White Game Bantam

In this variety also, the breeding practice is identical with that of the large corresponding variety (see page 274). The following defects must be guarded against in so far as possible: brassiness in males and white or blue legs.

While this variety formerly tended to be inferior in type, great improvement has been made in this respect, until now they are nearly, if not quite as good in type, as the other varieties.

The Black Game Bantam

This variety is bred in the same manner as the large variety. (See page 274.) In breeding, guard against the following defects in so far as possible: too wide feathers; purple barring; white lacing in hackle; too short legs; too much spread to tail and too long wings.

CHAPTER XIII

THE ORIENTAL CLASS

The Black Sumatra

In type, this fowl is a medium-sized bird of graceful shape and with a very long, low-carried tail which is abundantly furnished with sickles and coverts. See Fig. 93. In color, it is black, showing in both sexes a very high greenish sheen.

Fig. 93—Back Sumatra male showing long, well-furnished and low-carried tail. (Photograph from the Bureau of Animal Industry, United States Department of Agriculture.)

In making the mating for this breed, it is most common to use the single or standard mating, selecting birds of both sexes which approach as nearly as possible to the standard requirements. The male in this mating should be a highly colored, bottle green bird, while the females used should have a medium green sheen. In order to get the highest sheen in both males and females, a double mating is sometimes resorted to. To secure high-colored males, use a standard or exhibition-colored male with real dull-colored females. To secure high-colored females, use a high-colored male, which may show a little straw color in hackle and possibly in saddle, with high-colored bottle green females.

The Black-Breasted Red Malay

This breed is different in type from any other of the standard breeds. A very high station or reach is desired, and birds of both sexes should be selected which are fully up to or a little greater than standard in height, as there is a decided tendency for them to be too low. They should be decidedly higher than the Cornish. It is very important to have big, long bones and a good muscular development. The back should be broad and should slope from shoulders to tail to a marked degree. Avoid flat, level backs. The tail should be carried low, in fact, it should be inclined to droop and should just about carry out the incline of the back. The tail should be short and should be well folded together. Avoid any tendency toward high tail. The head should be broad, with the crown or skull projecting well over the eyes. Avoid birds for breeding which are too narrow and too long in head. The hackle should be as short and scanty as possible, and it is necessary to guard against a hackle which inclines to be too long and lacks trimness, that is, which is not neat in males. The legs should be set wide apart, and any tendency toward narrow-set legs or toward knock-kneed or cow-hocked legs must be avoided. As stated before, the bone of the leg should be coarse and heavy. There is a tendency for this breed to grow weak in hocks, the joints becoming flat and deformed if the birds are grown too fast. This breed should show an arched back, that is, it should show a convex curve from shoulder to base of tail as viewed from the side, and should not be flat or straight. This breed should show a distinct gullet, and it is necessary to select against birds which do not have this. The eye should be light or pearl in color, and as there is a tendency for the eye to come darker, birds with such dark eyes must be avoided. The comb desired is a strawberry or knob comb, but there is a tendency for the birds to show pea comb instead. A bird showing this character should not be used in breeding. It is necessary to be particular to keep the type of this breed distinct from that of the Cornish, to which it is somewhat inclined to run, due to a trace of Cornish blood in many strains.

In the male there is a tendency to get away from the reddish maroon color and to approach the light top coloring of the Black Reds. Avoid this and select for birds more on the cinnamon red. Avoid white in the under color of breast, body and fluff of males. Occasionally there will be some white in the under color of hackle and saddle of the male and in the under color of thighs. This, of course, must be avoided. There is also a tendency to show some red lacing in the breast and red frosting in the body, stern and legs of males. Breed from a male which approaches the standard in color as closely as possible, unless the females show a decided tendency to come too light in color, in which case breed from as dark a male as can be obtained. In the female there is a general tendency to come too light all over. The female top color is apt to approach wheaten color or to be too light a shade, while the breast may tend toward a salmon and the under color may run too light. With age hens tend to grow lighter in color, but if they were good in color as pullets they may be valuable as breeders.

The following is a brief resume of the defects which must be guarded against, in so far as possible, in making the mating: too low station; too light and too short in bone; insufficient muscular development; flat, level backs; too high, too long, or too well spread tail; too narrow and too long head; hackle in males too long and lacking in trimness; narrow-set legs; cow-hocked legs; knock-knees; weak, flat and deformed hocks; flat or straight back, showing no arch; lack of gullet; too dark eyes; pea comb; type like the Cornish; too light top color in males; white in under color of breast, body and fluff of males; white in under color of hackle, saddle or thighs of males; red lacing in

Fig. 94—Well-marked Black Breasted Red Malay Bantam feathers. M indicates male and F female. (Photograph from the Bureau of Animal Industry, United States Department of Agriculture.)

breast of males; red frosting in body, stern and legs of males; too light top color in females; salmon breast in females; too light under color in females.

The Black-Breasted Red Malay Bantam

This breed should be an exact counterpart of the large Black-Breasted Red Malay, the only difference being that of size. In color and in type exactly the same considerations (see page 281) must be observed in making the mating.

CHAPTER XIV

THE ORNAMENTAL BANTAM CLASS

The Sebright Bantam

The Sebright Bantam is perhaps the most exquisite of the standard varieties of chickens. It is of very small size, but perfectly formed and is extremely neat and dainty in appearance. It is also very sprightly. In the character of feathering and of coloring the two sexes are identical, the males being hen-feathered. The males are distinguished from the females by their slightly larger size, and the fact that they are somewhat coarser, particularly in the head and comb. This breed possesses a lacing in all the feathers of the plumage which is probably the most perfect lacing to be found in any breed.

In type, the birds are rather upstanding, with full, prominent breast, wings carried low, and tail well spread and carried high. This high carriage of tail gives the back the appearance of being short. The legs are rather short, and any tendency toward too much length of leg must be avoided.

The comb is rose, and should be rather broad in front and comparatively low and compact, with a well-developed spike, which tilts up slightly at the rear. Too high combs should be avoided, as where these are found there is a slight tendency for them to lop over on the side of the head. Combs should also be avoided in which the leader or spike is absent, as this is somewhat of a troublesome tendency in the comb of this breed. The comb should also be well filled out in the center, and should show no hollow at this point. The male comb in general seems to fill in very well, but the female comb is more inclined toward hollows in the center. Narrow combs are undesirable.

The hen-feathered males which usually are obtained in this breed are likely to give a very low percentage of fertile eggs. For this reason, it is well to mate only a small number of females to a male—about four. It also seems best to breed cockerels to hens rather than to use older males. A considerable percentage of the hen-feathered males will not fertilize eggs, and for this reason it is better to breed from males which have short sickles protruding above the main tail feathers at least one inch.

The following common defects must be guarded against in so far as possible in the matings of this breed: absence of leader or spike on comb; too high comb, which may occasionally lop over on side of head; hollow in center of comb, especially in females; narrow comb; too much length of leg.

The Golden Sebright Bantam

The color pattern of the Golden Sebright Bantam is the same as that of the Silver Sebright Bantam except that the ground color of the silver is replaced by golden. In making the mating, the same considerations in general hold true (see page 287). Golden Sebrights do not, however, show frostiness quite so much as the Silvers. The Goldens, which are darker in ground color than standard, are apt to be too heavily laced. It is also necessary in this variety to guard against any color which may range from light to white in the main wing and tail feathers, or any white in the quills of the primary feathers extending over one-half the length of the feather. Birds having this excessive amount of white in the quills have a tendency to produce chicks which show considerable white in the wing and tail, the wing especially.

In making the mating for this variety, it is necessary
to guard against the following defects in so far as possible, in addition to those which are common to the
breed (page 285): frosty lacing, especially on breast;
peppering or smut in tail; uneven lacing; lacing lacking
under throat and on head; too heavy lacing; peppering
on wing; light or white in main wing or tail feathers.

The Silver Sebright Bantam

While the lacing in both the Silver and Golden Sebright is perhaps the most perfect of any breed, still
there is a tendency for the lacing to be too wide or
heavy. There is also a tendency for the lacing to be
wider at the tip than at the sides of the feather. In
mating this variety, use a single or standard mating in
which the lacing throughout in both sexes is narrow and
as clearly defined as possible, and even in the different
sections. The lacing should be as free as possible from
any frostiness, this frosty lacing being most likely to
show on the breast. Frosty lacing is a very bad fault,
and must be carefully avoided. Another important point
is to be sure that the lacing is even over the entire bird.
Birds which show good lacing on the breast are likely
to prove good breeders, for this is one of the most difficult sections to secure good lacing. Birds which lack
the lacing under the throat and on the head should not
be used for breeding. Frostiness appears more commonly on the breast than on any other section. The
narrower laced the birds are, the greater is the tendency
for the tails and wings to come free from smut or black.
Select birds which are as free as possible from black peppering in the main wing and tail feathers. This peppering is especially troublesome in the wing. Any dark or
nearly black color should be avoided in the wing and
tail. Most birds have some of this, as birds entirely free

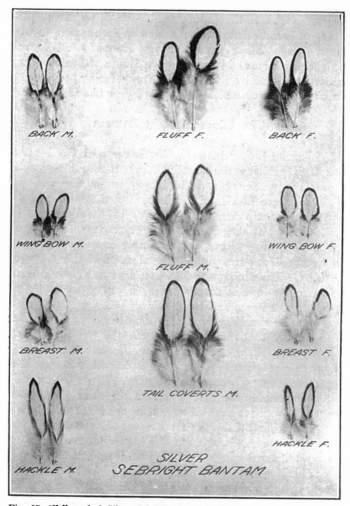

Fig. 95—Well-marked Silver Sebright Bantam feathers. M indicates male and F female. (Photograph from the Bureau of Animal Industry, United States Department of Agriculture.)

from it are very rare. Where it is necessary to use females which are too heavily laced, it is best to use a light-laced male, as such a mating is most likely to get the proper lacing. The reverse of this mating also holds true.

In making the mating in this variety, it is important to guard against the following defects in so far as possible, in addition to those which are common to the breed (page 285): frosty lacing, especially in breast; peppering or smut in tail; uneven lacing; lacing lacking under throat and on head; peppering on wing; too heavy lacing.

The Rose Comb Bantam

The Rose Comb Bantam is a very beautifully molded little bird, and in type should be practically a miniature of the Hamburg, except that the tails and the wings are considerably larger in proportion to the size of the bird, and that the wings are carried low. The tails are also carried low, and any tendency toward high tail in either sex should be avoided.

The comb of the Rose Comb Bantam should be of fair size, so as to be in proportion to the bird. Any tendency toward too small a comb, especially in the female, must be avoided. In shape, the comb is a little more rounded than the Sebright comb, being just about like the Hamburg. The spike inclines upward slightly and should never show any tendency to follow the neck. Some combs give the appearance of being hollowed out on the top in front. Really, however, this is more the result of the outer edges of the comb being too high rather than the center being hollowed out.

The ear lobe should be large, white and rounded, and should show no tendency to be hollowed out or to be concave in character. This is especially likely to occur in the males. Hollowed-out lobes have a tendency to show red around the bottom. The lobe should be full and smooth.

The face should be a bright red in color. In the males there is some tendency toward white in face, but this is more likely to develop with age, and does not occur so frequently in young birds. It is a more serious fault in young birds than where it develops with age in the older birds. There is also some tendency for the face to come gypsy colored, in the female especially. Matings are sometimes made in which gypsy-faced females are used in order to get rid of any tendency toward white in face.

There is also somewhat of a tendency for the males to show narrow sickle feathers. It is necessary to select birds which have good width of sickle feathers and tail coverts.

In selecting the matings in this breed, it is necessary to guard against the following defects in so far as possible: too high tail; too small comb, especially in the female; spike following the neck or not inclining upward; combs hollowed out on top; outer edges of comb too high; hollowed or concave ear lobe; red in ear lobe; white in face; gypsy colored face, especially in females; too narrow sickle feathers in males.

The Rose Comb White Bantam

In this variety a single or standard mating is employed. The color of the breeders should be a pure white throughout, and should be free from any tendency toward brassiness, especially in the plumage of the male. Considerable black ticking is apt to occur in the white plumage, and this is more prominent in the tail and flight feathers. This, of course, must be guarded against. Birds sometimes have blue instead of white legs, and these should not be used in breeding if it can be avoided. For defects common to the breed which must be guarded against in so far as possible, see page 289.

The Rose Comb Black Bantam

This variety is generally considered to be a trifle more hardy than the Rose Comb White Bantam. In making a mating, it is usual to mate a standard male and female. Some breeders advocate putting in a few larger comb females in the matings, so as to offset any tendency for the female to come with too small a comb.

The color should be a beautiful, lustrous, greenish black throughout the surface of both sexes. It is quite a common practice to use a high-colored male with dull colored females, or vice versa, to get this green sheen and to get rid of purple. Males are also obtained occasionally which show some straw color in their hackle. This occurs as a dark stripe in the hackle feathers edged with straw color. These males mated to high-colored females which are as near beetle green as possible will give high-colored females. In the Black Rose Comb Bantam purple in the plumage, both as a tinge and as a barring, is quite troublesome, and it is necessary to select against this in the mating. There is also a slight tendency in this variety toward gray eyes. Another very bad defect and one which must be carefully guarded against is to see that the breeders of both sexes show no gray in the wing feathers.

In making the mating for this variety, it is necessary to guard against the following defects in so far as possible, in addition to those which are common to the breed (page 289): purple tinge or purple barring; straw in hackle of males, except for special matings; gray eyes; gray in wings.

The Booted White Bantam

This breed should be well up on its legs, with a back which is short, but not extremely so. The tail should be well spread and carried moderately high. There should be no angle between the tail and the back, these two sections joining smoothly with a nice curve. It should be heavily booted or feathered on the shanks and on the outer and middle toes. It is especially important to guard against birds which have a bare middle toe. The birds should have a vulture hock. It will be seen that with reference to the bird being well up on legs and possessing stiff vulture hocks, it is in these respects just the reverse of the Cochin. The comb, which is single, should be of medium size, and it is necessary to guard against too large combs, which are most likely to be found on males. The color of plumage should be a pure white throughout, and the color defects to be guarded against are brassiness and dark or black ticking.

The Brahma Bantam

The Brahma Bantam in type should be as nearly as possible an exact miniature of the full size Brahma. There is a decided tendency for the Brahma Bantams to run too large in size, and it is necessary to select carefully against the larger birds to offset this tendency. There is also a considerable tendency for the males, in particular, to have vulture hocks. This must, of course, be guarded against, for while it does not constitute a disqualification, it is a grave defect. The comb should be a typical pea, but there is a considerable tendency for the middle blade of this comb to be too high. As in the regulation Brahma, the middle, as well as the outer, toe should be feathered, and any tendency toward a bare middle toe must be selected against. For further information as to the desired type and defects which must be guarded against in so far as possible, see the material under the Brahma, page 131.

The Light Brahma Bantam

In breeding this variety, both the single or standard mating and double mating systems are used. In general, the same matings are made in this variety as are made in the corresponding variety of the regulation Brahma (page 134). For information on the double matings used, see the material on the Columbian Wyandotte, page 114.

It is necessary to guard against brassiness in the white of the male plumage. There is also a tendency for the under color of hackle to run too light in both sexes. Birds which possess this light under color in hackle are apt to lack the lustrous black in hackle, and to have white shafts to the feathers, causing a gray appearing hackle. Many females show too much black ticking in back. This is likely to be associated with too dark an under color of back. In general, the birds of this variety show good black wings, but many of the birds which are good in the black of wings are apt to show ticking in the back. For defects common to the breed which must be guarded against in so far as possible, see page 292.

The Dark Brahma Bantam

It is usual in mating this variety to resort to the double mating system. In general, the same matings are made in this variety as are made in the corresponding variety of the regulation Brahma (page 140).

The principal color defects which must be guarded against are brown in the lower edge of the wing-bow of males and females, which run stippled over the back. This back stippling of females is a serious defect, which is especially likely to occur in the Dark Brahma Bantam. For defects common to the breed which must be guarded against in so far as possible, see page 292.

The Cochin Bantam

The Cochin Bantam should be in type an exact duplicate of the large Cochin (page 146), but, of course, small
in size.   There is a decided tendency for all varieties of
the Cochin Bantams to come too large in size.   The
Buffs have probably been bred so that they run more
uniformly small in size than the other varieties.   There
is also a tendency for all the varieties to be too high on
legs, especially in males, and this must be guarded
against.   The comb, which is single, should be medium
in size, and should be strongly erect.   All the varieties,
however, have a tendency to run too large in comb, the
Buffs and Blacks being the worst in this respect.   With
a large comb, which is apt to occur, there is likewise a
tendency toward lopped combs.   It is well to select birds
which have a small comb and in which the combs are
relatively thick at the base and are strong, that is, not
too thin.   Be sure to select the loosely feathered birds
for breeders, as there is somewhat of a tendency for the
birds to be short and close-feathered.   The White variety is the worst in this respect.

As in the large Cochins, there is a considerable tendency for the birds to have vulture hocks, and since this
is a serious defect, it must be carefully selected against.
The White variety, perhaps, shows the worst tendency
in this respect, while the Buff is the least likely to show
vulture hocks.

In selecting the matings for this breed, it is necessary
to guard against the following defects in so far as possible: too short or too close feathered; too high on legs,
especially males; too large comb; lopped comb; vulture
hocks; too large in size.

The Buff Cochin Bantam

In selecting the mating for this variety, both the single
or standard and the double mating systems may be used.

In the single mating, use the male which has an even color all over; use good, golden buff females. If pullets are used they must be standard in color, but if hens are used they may be one shade lighter, but should by no means be a washy, pale buff. Eliminate any males from the breeding pen which show a red or which run to a lemon; also guard against males which show a stronger colored wing bar. In under color, the shaft of the feathers must be buff to the skin; while a buff in the rest of the feather is desired, it is not so important as that the shaft be buff. The light shaft will show in the surface of the female backs and will produce shafted females. One of the greatest defects to guard against is white in the wings of the males. Be sure that all the females in the breeding pen have strong wing color.

Where the double mating is used, select as follows:

Cockerel mating.—Use the standard male or one which tends to be a trifle lighter in color than the standard, as there is a tendency for the males to run too dark. The females selected should run a trifle lighter than the male bird's breast, but must have strong color of tail, wings and legs.

Pullet mating.—Use a rich-colored male which is a little darker than standard. Some breeders recommend a male in this mating which shows a trifle heavier color in the shoulders than the standard, but the shoulders should be without any sign of red tinge. Use good, golden buff females whose surface color or ground color should match the male bird's breast.

In making any of these matings, it is necessary in so far as possible to guard against a wing-bow which is darker than the breast, especially in the male; unevenness in surface color; white in wings and tail, or dark in tail; white in under color (birds showing white under color sometimes have white in wings); a tendency for saddle and tail coverts to be laced with white, but good color

otherwise; eyes which are lighter than the reddish bay
called for by the standard.   Additional and more detailed
information on breeding for buff color will be found
under the Buff Plymouth Rock, page 89.   For defects
common to the breed which must be guarded against in
so far as possible, see page 294.

The Partridge Cochin Bantam

In selecting the matings for this variety, it is usual to
employ the double mating system.   The general considerations in selecting these matings are exactly the same
as in the case of the large Cochins (page 147).   The males
in this variety generally come pretty good in top color,
the main weakness being that the striping in the hackle
and saddle are not well defined.   There is also quite a
decided tendency in this variety for birds of both sexes
to run too large, and the effort should be to breed them
as small as the Buff Cochin Bantam.   There is also a
tendency for the females to be too light in ground color.
This ground color should be a rich mahogany.   The females also tend toward poor penciling, that is, penciling
which is not well defined, but which more nearly approaches stippling.   They are especially apt to be stippled
over the back and to be too light colored in throat and
breast.   For defects common to the breed which must be
guarded against in so far as possible, see page 294.

The White Cochin Bantam

This variety has a greater tendency than any of the
others to be short and closely feathered.   It is essential,
therefore, that breeders be selected which are as loosely
feathered as possible.   In selecting the matings, the
single or standard mating is employed and birds should
be picked out which approach as closely as possible to
the standard. Select birds which are pure white through-

out and which are free from any signs of brassiness. It is also necessary in this variety to select against those birds which have a tendency to show long, stiff tails. This is a defect which is more or less common, and which is undoubtedly due to the Booted White Bantam blood which has been used in making the White Cochin Bantams. For defects common to the breed which must be guarded against in so far as possible, see page 294.

The Black Cochin Bantam

This variety is probably the most typical of all Cochin Bantams, as it approaches more nearly the shape of the large Cochins. It is usual to employ a single or standard mating in which birds of both sexes are selected which approach as nearly as possible the standard. The black surface color should show a greenish sheen and should be free from any purple tinge or purple barring. There is also a tendency for the birds to come with dark eyes, and this should be selected against, as red eyes are desired. White also may occur in under color of hackle and saddle and this, of course, must be selected against. For defects common to the breed which must be guarded against in so far as possible, see page 294.

The Japanese Bantam

This breed is quite unique among the standard breeds of poultry. It has a fairly large single comb which is erect in both sexes. On account of the large size of the comb, there is a considerable tendency toward lopped comb. The neck is comparatively short and the body very low set, the shanks being extremely short. The tail is large and well spread, and is carried so high that it is decidedly squirrel; in fact, the tail should almost touch the neck. To secure birds having tails carried well forward, it is well to breed from birds in which the

tail and neck touch, if possible to secure them. The
wings are large and are carried very low so as almost to
drag upon the ground. This breed should be very heavily
feathered, but should be compact in feathering so that it
is not Cochiny in that respect.

In making the matings in this breed, the following defects, which are common to all varieties, must be guarded
against in so far as possible: wry tails; narrow feathers
in the tail and soft texture of feathers; too much length
of leg; stubs; too loosely feathered birds; lopped combs.

The Black Tailed Japanese Bantam

In making the mating in this variety, it is common to
use the single or standard mating, selecting birds of both
sexes which are as nearly like the standard as possible.
There is some tendency to have white in the main tail
feathers, this being especially true of the females.
Breeders should be selected in which the main tail
feathers are black. It is also necessary to guard against
brassiness in the surface color of males, and against
black ticking in the neck hackles of both sexes. Dark
in the under color of back is also undesirable, as it is apt
to be associated with ticking in the hackle. For defects
common to the breed which must be guarded against in
so far as possible, see page 298.

The White Japanese Bantam

In mating this variety, the single or standard mating
is used. Birds of this variety are inclined to run a trifle
higher on legs than other varieties, and this must be
offset in the breeding by the selection of birds low on
legs. It is also necessary to guard against brassiness in
the surface color. There is likewise some tendency
toward ticking and a tendency to show black in the tail.
For defects common to the breed which must be selected
against in so far as possible, see page 298.

The Black Japanese Bantam

In mating this variety, use the single or standard mating. The principal color tendencies which must be guarded against in so far as possible are gray in the primary wing feathers, white in the under color in all sections, and any purple tinge or barring, especially in the males. It is hard to get birds which are absolutely free from purple. For defects common to the breed, see page 298.

The Gray Japanese Bantam

In mating this variety, use the single or standard mating. The Gray Japanese Bantams should have Birchen markings. The males come better in this respect than the females, as the latter are apt to come with stippling on back and saddle, and on the tail feathers and secondary wing feathers. The females also show a tendency to come dull black, with poor lacing; that is, the lacing on the breast is inclined not to be clear and distinct, the black tending to run into the white. The hackle is also inclined to be mossy, that is, the white running into the black centers. Guard also against gray in the fluff of females. It must be remembered, however, in selecting females for the mating that it is difficult to get females which are good in color in this section. The males come pretty good in color of fluff. This variety is inclined to run more Cochiny in feathering than the other varieties. This, of course, must be selected against. For defects common to the breed which must be guarded against in so far as possible, see page 298.

The Polish Bantam

The Polish Bantams, both in the bearded and non-bearded varieties, are identical with the large Polish (page 229) except in the matter of size. There are, however, several tendencies which should be mentioned in connection with the Polish Bantam. These fowls tend to have too much comb, the comb approaching the leaf comb in character. It should be very small, appearing merely as two little horns. The males also have a tendency to have short sickle feathers and to lack long, flowing tail coverts. Only a few males have good length of sickles and long flowing coverts.

The Bearded White Polish Bantam

In mating this variety, exactly the same consideration should be given as in the case of the large White Polish (page 235), but it is also necessary to guard against a blue or black face. While some judges seem to prefer a blue face, this is wrong. It is also necessary to guard against the closed or low nostril. Breed from birds which have open or raised nostrils (page 253), just as in the big variety, except that they are smaller. This variety is very subject to closed or low nostril. The skin should not be blue, but the majority of the birds have a blue skin. If possible, breed from those with white skin. For defects common to the breed, see page 300.

The Buff Laced Polish Bantam

In breeding this variety, the same matings are made as in the case of the large Buff Laced Polish (page 237). There is a tendency for birds of this variety to have white beaks, and this must be guarded against. There is also a considerable tendency for white in the tails and wings, which must also be selected against. For defects common to the breed, see page 300.

Non-Bearded Polish Bantams

In mating the non-bearded varieties, it must be kept in mind that they are the same in every respect as the corresponding bearded varieties, except that they are without beard. In making the matings, the same consideration should be given as in the case of the large Polish.

The Mille Fleur Booted Bantam

The Mille Fleur Booted Bantam is really a variety of the booted bantam, and should not be considered a separate breed. The Mille Fleur simply refers to the color and color pattern. In type, therefore, the Mille Fleur Booted Bantam is identical with the Booted White Bantam (page 291).

In making the mating, it is usual to employ a single or standard mating. Avoid females showing shafting and stippling in the buff ground color. This is most likely to occur on the back. The buff color should be clear. Specimens should be selected which have each color distinctly separated. The white tip should not be too large, and the black bar which separates the white tip from the buff part of the feather should also be of moderate size. There is some tendency for the white tip to run into the black if the tip is too large. This tendency increases with the age of the bird. Pullets and cockerels in their first plumage should not show a great amount of white tipping on account of this increase with age or after the molt. Avoid males as breeders that have striping in the neck and saddle, running through the entire length of feathers. These males are apt to produce females which show shafting and stippling. Avoid solid white flight and tail feathers in both sexes. The boots should be as heavy as possible, and the hock feathering long and stiff. There is a tendency for the booting to run too light.

The correct shade of ground color in the female should not be a pale lemon or deep bay, but should be a rich, golden buff. In pullets a shade slightly darker than this may be tolerated, as it tends to lighten with the molt,

The proper shade of under color in the female should be a bluish slate next to the surface, shading into a pale salmon color toward the skin. If the entire under color is slate it is apt to be associated with too heavy a black bar

Fig. 96—Well-marked Mille Fleur Bantam feathers. M indicates male and F female. (Photograph from the Bureau of Animal Industry, United States Department of Agriculture.)

separating the white tip from the buff color. Where the under color is all buff, it is apt to be associated with a too scanty black bar. In case it is necessary to use a male bird which is dark in ground color, he should be mated to light-colored females.

The males of this breed tend to have a large comb. These bantams do not, as a rule, take on their correct plumage until they are two years old. Therefore, it is best not to cull or dispose of young stock until the end of the second year. Avoid any solid colored body feathers; that is, those lacking the spangle and the bar, and also avoid any solid white or black feathers in the hackle of both sexes. There is somewhat of a tendency for the shanks to come yellow, and this must be avoided, as the correct color is slaty blue. The Mille Fleur breeds both plain and bearded.

CHAPTER XV

THE MISCELLANEOUS CLASS

The Silkies

As the name implies, this breed is characterized by the peculiar nature of its feathering. The feathers should be long and webless and of a silky texture. The breed is of small size, being practically Bantams. In type they should be low, resembling somewhat the Cochin Bantams. The feathering is profuse. They possess a crest and a fifth toe. The shanks and outer toes are feathered. The comb should be small and round, without any spike. There is, however, a tendency for the comb to have a single leader or spike, or in some cases to have a small double spike. The face and comb should be purple in color, and it is necessary to guard against red or white lobes and red combs and faces. The ear lobe should be turquoise blue. Good lobe color, face and comb color are among the hardest things to get right in this breed. The feathers of the wing and tail sometimes tend to be stiff, but this should be avoided. There is somewhat of a tendency for the fifth toe to be set close to the fourth toe. This is undesirable, as the two toes should be well separated. Birds with four toes must also be guarded against, as the absence of the fifth toe is undesirable and quite a common defect. Occasionally six toes occur, and this must also be avoided. Any tendency toward too great a length of leg, which makes the fowl set too high on leg, must be avoided. Avoid stiff quill feathers around the hocks, as there is a tendency in this breed toward vulture hocks, which are undesirable. The crest tends to run too small in size, but should be full and of moderate size.

The hens are splendid sitters and mothers, and are extensively used in hatching small sittings of choice eggs of other varieties and the eggs of pheasants. They are naturally quite tame.

In making the matings of this breed, it is customary to employ the single or standard mating, selecting birds which approach as closely as possible to the standard in both sexes. The plumage should be pure white throughout, and there is little tendency toward brassiness.

Avoid the following defects, in so far as possible, in the birds to be bred: too much length of leg; shanks and outer toes not feathered; red or white ear lobes; red comb; red face; stiff web and quills of wing or tail feathers; absence of fifth toe; fifth toe not well separated from fourth; the presence of six toes; too small crest; vulture hocks; comb with leader or spikes, or with small double spike; color of leg, skin and beak other than blue.

The Sultans

This breed is medium to small in size and characterized by a large crest, the presence of a muff and beard and vulture hocks. In making the matings it is customary to use the single or standard mating, selecting birds of both sexes as breeders which approach the standard as closely as possible. The plumage should be pure white throughout, and free from any foreign color.

In this breed a good but small V-shaped comb is desired, and a beard which is the same as that of the bearded Polish. The shank and foot feathering should be heavy, like that in the Booted Bantams. An open or raised nostril like that in the Polish (page 229) is desired.

In making the matings it is necessary to guard against the following defects in so far as possible: lack of vulture hocks; lack of fifth toes; lack of beard; too scanty foot feathering; feathering not extending down the shanks and over the outer and middle toes; closed or single nostril (birds with closed nostril have a tendency to throw single combs); brassiness.

The Frizzles

The birds of this breed are characterized by a peculiar character of the feathers, which show a decided tendency to curve backward or upward at the ends. This is especially noticeable in the hackle and saddle feathers, but it is desired that all of the feathers show this to the greatest degree possible. The comb should be single and the color, which must be solid, may be either black, white, red or bay. When these birds are shown in pairs or pens, it is necessary that they match in color. They should have four toes. So far as general type and other characteristics are concerned, the standard is not well fixed. As soft feathers as possible are desired on the males, and it is necessary to guard against stiff wing feathers. It is also necessary to guard against short-feathered birds, as these do not show the frizzled character. Frizzled males, when bred to long, soft-feathered females with a single comb, such as the White Rock, will produce about 50 per cent of the offspring frizzled.

CHAPTER XVI

PREPARING FOWLS FOR THE SHOW

The successful exhibitor overlooks no opportunity to show his birds to the best possible advantage. He realizes that competition is keen, that the other exhibitors will do everything possible to put their entries in shape to win, and that very small differences in condition may be sufficient to make a bird a winner or a loser. He begins, therefore, in a sense, to prepare his fowls for the show from the time that they are hatched. The birds must be given every chance to show to the best advantage. Failure to condition and prepare birds properly may result in a well-conditioned bird of fairly good quality winning over a better bird not properly conditioned. The breeder must therefore give careful attention to the matter of preparing his fowls for the show.

Time of hatching.—The time of hatching is important. There is only a comparatively short time in a young bird's life when it is at its best, and this is when it has just reached full maturity, full growth of body and feathers, and while its plumage is still fresh and in the bloom. Pullets show this condition just as or just before they begin to lay. After they have laid for a time their freshness and bloom are lost to some extent, and they do not appear to as good advantage. It is essential, therefore, in order to have the young stock at its best, that it be hatched at the right time. For the earlier fall shows the hatching should be done early, probably in February or even January, while for the later shows the hatching may be later. It seldom pays, however, to hatch stock for exhibition after the middle of May, for this does not as a rule allow time enough for the development and growth of birds of standard size. Later hatched birds may prove to be suitable for the late winter shows and to show as hens and cocks the following year, but since there is a tendency in nearly every breed for fowls to go to pieces, or, in other words, to develop defects with age which were not present when the birds were young, it is better to hatch early enough so that the fowls can be used for exhibition purposes as cockerels and pullets if of the proper quality, and to depend upon such young birds for exhibition cocks and hens in succeeding years. In some breeds there is a decided tendency for the birds to fall below the standard size and weight. In such cases early hatching is especially important, in order to give the fowls every chance to reach as large a size as possible.

Feeding and management of growing stock.—If young stock is to have the size and development necessary in order to be of exhibition quality, it is essential that it be well grown. The growth of the birds must have been continuous and fairly rapid, with no serious checks or setbacks. Any method of feeding growing chickens and any rations which have been used and which have given good, continuous growth are satisfactory. In such rations a considerable variety of feed is desirable in order to insure the growth of a strong, rugged frame and luxuriant plumage. If young birds seem to be developing too rapidly, they can often be delayed by changing the feed and keeping them almost wholly on a grain mixture. Pullets can often be prevented from laying by moving them to strange quarters.

As the fowls approach maturity, care must be exercised not too feed too much beef scrap. This is a stimulating feed and may cause the combs of cockerels and pullets to get too large or to get large too quickly. It may also cause the pullets to start laying at an early age, which not only tends to stunt their growth, but is also undesirable, because it will destroy the freshness and bloom of their appearance when shown.

As soon as the cockerels begin to annoy the pullets to any great extent, they should be separated from them and placed in separate yards or on a separate range. In removing the cockerels, take away those of about the same size and development. As others develop, they can be removed later, but should not be placed with those taken away previously or they will fight and injure one another. If the cockerels are left on the pullet range in considerable numbers, they will annoy them so much as to interfere to some extent with the growth of the pullets. A few promising cockerels which are backward in their development may be left on the pullet range to good advantage, as they will have a better chance to develop than if placed with larger cockerels, which will cow and dominate them. When any exceptionally outstanding cockerels are noticed, it is best to separate them from the others, either by placing them in a pen by themselves, or, preferably, by putting them with a small pen of hens. If left with the other cockerels there is always a danger, even though they have been raised together, that they will fight and be permanently injured as show birds.

A cockerel which is backward in development can often be brought along more quickly by placing him with a few hens than if left alone or with other cockerels.

Feeding white fowls.—In white fowls, or those with large white sections, there is sometimes a tendency for the white of the plumage to be creamy or brassy. While this is a matter of breeding to a great degree, the feeding will increase or decrease the amount of yellow in the plumage according to the materials used. This is true not only of young fowls, but of older fowls as well, particularly just preceding and during the molt. For white fowls, never use yellow corn. If corn is used, purchase and feed white corn. Avoid also any feeds which are rich in fats or oils, such as sunflower seed, oil meal, cottonseed meal, peanut meal, etc., and also any form of meat feed which contains much fat. When it is desired to feed meat of any kind, use that as free from fat as possible, and it may be found advisable to boil the meat before feeding in order to extract the fat. Hog liver is said to be free from the yellow pigment which causes this creaminess, and is used by some breeders in feeding their white fowls. Green feed is also bad to bring out creaminess in white fowls. Of course it must be kept in mind that if one should go too far in cutting out the use of feeds which may affect the color of white plumage, they would be likely to get an undesirable effect in another direction, in the case of yellow-legged varieties, in a partial loss of leg and beak color. It should also be noted that the same feeds which bring out creaminess in white plumage will help to bring out creaminess in white ear lobes.

Value of shade.—Shade is of the utmost value, if not absolutely essential, in obtaining the best color of plumage. Intense sun has a bad effect upon practically any color of plumage. It may cause sunburn or brassiness in white birds, purple bars in black, and will fade or deaden buff, red, barred Plymouth Rock or any other colored plumage. The combined effect of the sun and rain is worse than that of the sun alone. It is frequently claimed that the color of late-hatched is better than that of early-hatched birds. This is undoubtedly due to the fact that the early-hatched fowls get their mature plumage earlier in the year than the late-hatched, and in consequence it is exposed for a longer time and to a more intense sunshine, which causes a greater degree of fading. A note of caution may be needed here, with respect to brassiness in white fowls. While this color may be brought out by exposure to the hot sun, there is a great difference in strains in respect to the occurrence of brassiness under any conditions, some being much freer than others. Therefore the breeder should not fool himself, if he has birds free from this color due to the fact that he has kept them out of the sun, into thinking that his birds are free from this defect from a breeding standpoint.

Selecting the birds to be prepared for the show.—The successful breeder will prove to be a student of his birds from the time they are hatched. He watches them closely as they develop, and will have in mind those which show the most promise and which stand out as exceptional birds. His choice of the birds to be conditioned will therefore be among those birds which have attracted his attention. Needless to say, only those birds which are in good health and natural condition would be among them. The next step in their selection will be a close examination, section by section, looking especially for serious defects which will put them out of the running, and for strength and weakness in the different sections and for the fowl as a whole. The condition and development of the birds, with respect to the time of the show, will also be a big factor in their selection. Often birds can be taken direct from the range or from the pens in about the best possible condition in which it is possible to get them, except for washing and other special attention. The breeder should select for conditioning several birds more than he expects to exhibit, as some of them may go bad during the process.

It is well to group the birds showing promise together in exhibition cages, so that they can be compared side by side, as this affords the best possible opportunity to study them and to select the best. The breeder will already have a good idea of the excellence of his older birds, but they must be carefully examined to see that they have not gone bad in any section. The choice will then lie with those which are in the best condition of health and feather for the particular show.

It sometimes happens that the judge which is to officiate tends to emphasize type rather heavily in his work, or color, or pays especial attention to some particular quality or section. If the breeder knows the judge's taste and preference, he will frequently make selection of certain birds which are exceptionally strong in these particulars, even if not in his judgment quite such good all-around birds.

Cooping birds for training.—When it becomes time to train the birds, they should be cooped separately in coops which are as nearly as possible like those used in the show room. Be sure that the coops are large enough so as to give the birds plenty of room to stand in a natural position. The usual exhibition coop is 2 feet long, 2 feet deep and 27 inches high, and the coops used for training should be about this size. Also be sure to use coops with open wire tops, so that the birds will become used to them and not try to fly out, as this will often injure the comb. If the birds show a tendency to try to fly out the top, cover it with a piece of paper or some other material until they get used to it. When a pen is to be shown, the females should be cooped together for training and before washing, as this will prevent the fighting which might occur if the hens were first put together in the show room. The usual pen exhibition coop is 4 feet long, 30 inches deep, and 30 inches high.

Training.—Remember that other things being equal, the best trained bird will win. Training is therefore important. Place the coops, if possible, where people pass by frequently, so that they will become accustomed in some degree to the condition which they will find in the show room. Some conditioners even train their birds in a room of the dwelling house, where someone is present most of the time, or the children are playing about, so that they will become used to the noise and motion. Never scare the birds, but move carefully and quietly; otherwise they may injure themselves or their plumage in their efforts to escape, and may be made wild and harder to train. Handle the birds whenever possible, and take them out of the cage so that they will be used to this when judged. Considerable time can be spent to advantage in posing the birds in the cage, both with the hands and with the judging stick, so that they will become used to either method. The attempt should be made to train the bird to assume a pose which will show him off to the best advantage whenever anyone approaches the cage. A great difference will be found in different birds as to the amount of time necessary to tame and train them. Some birds seem to be natural born show birds, and are relatively tame and instinctively assume a good pose. Other birds are harder to tame, and occasional birds can never be satisfactorily tamed and trained. If a bird is inclined to be wild, feeding sparingly will help to tame him. It will also be found helpful to take the bird out of the cage and carry him around on one's arm whenever possible. Some birds may be tamed and trained in two or three days, others will take a week, and others will need two weeks or longer. In general, it is easier to tame a cockerel or a pullet than a cock or hen which has never had previous handling. A cockerel which has been handled, if he received no further attention for a year, will show the effects of the previous handling when the attempt is made to train him as a cock. He will not have forgotten entirely, and will not be as difficult to handle. The room for training should be unheated, so that the fowls will have about the same temperature that they are used to and will be less likely to catch cold when taken out to ship. A heated room is also dangerous to use because it is likely to cause too great a comb development.

Danger of overcooping.—Do not keep the fowls too closely confined to the coops while training. Arrange, if possible, so that they can be given the run of a pen part of the time, as otherwise they will not have exercise enough and may get out of condition. Feed all cooped birds lightly, unless they are being fed for special development, as there is usually a tendency to overfeed. It should be mentioned here also that birds should not be shown too much. Usually it is best to show a bird at only one, or at best two shows. Not only is it next to impossible to keep the birds in the pink of condition, but the hardship and travel are such a trying experience for the bird that his health is likely to suffer as a result. This is especially true with respect to birds which are to be used as breeders, for the fertility of both males and females is very likely to suffer. If a bird is to be shown at two shows, he should be given a rest between, if possible.

Feeding birds which are being conditioned.—When birds are smaller than they should be, or too thin, it is sometimes possible by special feeding to increase their weights and bring them up to standard. To accomplish this, feed them plentifully of some good fattening ration, using a considerable proportion of moist mash. The greatest care and judgment must be used, however, to see that in the eagerness to put on weight the birds are not crowded too hard and do not go off their feed, and it is important not to coop too closely so that exercise is denied the birds. Where birds in training are not being fed for development, it is usual to feed only good, sound, hard grain, and to feed that rather sparingly. This prevents looseness of the bowels and tends to keep the fowls in good condition. A small amount of green feed should be given and a small amount of beef scrap or other meat feed, if the birds have been used to it. The latter must be fed very sparingly, however, as it is apt to overdevelop the comb, and it may be necessary to omit it entirely. Some conditioners feed soft feed in whole or in part in place of the hard feed. As a rule, this is practiced in the breeds where fluffiness of feathering is desired, as in the Cochin and Orpington. Nothing but hard feed is given to Games, where closeness and hardness of feathering are desired. Where a gloss of feathers is desired, it is common practice to feed materials rich in oil and fat, such as sunflower seed, oil meal, or meat feed containing fat, but this class of feeds should not be given to white fowls, especially when growing new feathers, as it brings out creaminess.

Cleaning shanks and toes.—The shanks and toes of all birds should be washed. In the case of birds whose plumage is to be washed, this should be done before washing the feathers, and in separate water so as not to dirty it. Use warm water and soap, and scrub the shanks and toes with a nailbrush until all dirt is removed. After this, take a piece of hard wood, sharpened to a point, or a toothpick, and clean out all the dirt which has worked under the scales. After the shanks and toes have been cleaned, the fowl must be kept in a pen or coop with clean litter, and not allowed to run on the dirt. Birds with yellow shanks should be kept off of unsodded heavy clay land if possible, and also away from coal ashes, as both these have a bleaching effect and are likely to cause the fowls to have too light-colored shanks.

Washing.—All white birds, or birds having much white in their plumage, should be washed. This is necessary in order to clean the plumage and is beneficial also because of the fluffing effect which it has on the feathers, which improves the appearance of the bird. Because of this fluffing effect and the brightening of the colors, many breeders wash colored as well as white birds. Other breeders of colored varieties feel that by giving the birds a conditioning pen, where they can exercise in plenty of clean straw, not too long, they will clean themselves satisfactorily and secure a natural bloom to the plumage, which is superior to washed plumage. Games, in which closeness and hardness of feather are desired, are, of course, never washed.

Washing should be done after training is practically completed, as the birds should be handled as little as possible after they are washed. Washing should be done two or three days before the birds are shipped. Use rain water or snow water, if possible, as it is preferable to hard water. Many hard waters also contain iron or some other stain which will have a disastrous effect upon the plumage. If hard water, known to be free from stain is to be used, it may be softened with borax or ammonia. The washing should be done in a room, the temperature of which is about 70 to 75 degrees. Ordinary washtubs are suitable for washing the birds, as they are roomy. Three or four tubs should be used. They should be filled about two-thirds full of water which is about 103 degrees in temperature. Work up a good lather in the water, using a sponge for this purpose, then, holding the bird with the legs in the left hand and with the right hand on the back, across the wings, lower it into the water. Sponge off the head, wattles and comb, so as to remove all dirt. Gently raise and lower the bird in the water until the plumage is thoroughly wet, then lather with castile or other good, pure soap. Wetting all parts of the plumage thoroughly before washing is of the utmost importance. A thorough soaking is necessary to accomplish this, and frequently when it is believed that the wetting is complete, dry parts of the feathers next the body will be found on opening up the plumage. Failure to wet the plumage thoroughly before washing is likely to result in breaking or injuring the feathers. When the bird is well lathered, the feathers may be worked over very thoroughly, always working the hands with the feathers and not against them. In this way the feathers may be thoroughly washed without danger of injury. The lather should then be washed out and the lathering repeated a couple of times, or until the plumage is clean. After the last lathering, the soap should be washed out as well as possible and the bird then transferred to tub No. 2. Here the feathers should be worked over again in an effort to get the soap thoroughly out. The bird is then transferred to tub No. 3, where it is carefully rinsed again. It is necessary to make sure that all of the soap is out of the plumage before the bird is taken out of this water, and it is well to have a fourth tub in which the bird can again be rinsed to make sure. If any soap is left in the feathers, they will not dry out right, but tend to be sticky and will not fluff out well.

Remove the bird from the rinsing tub, letting the water run off and pressing out as much as possible by running the hand over the surface of the feathers. A turkish towel should then be used to blot up moisture from the bird, but never rub the feathers with the towel, as this may break or injure them. The bird is next placed in the dripping cage. This is a cage with a wire bottom to allow the water to drip away. Needless to say, this cage and any others used after the bird is washed must be perfectly clean. The cage should also be large enough to allow the bird to shake himself and to flap his wings, as this will hasten the drying process. A stove with a brisk fire is necessary in the drying room, and the dripping cage should be placed near enough to the fire so that the bird is in a temperature of about 90 degrees.

After the dripping has ceased, remove the bird to a cage with clean shavings in the bottom. Over the shavings should be placed a paper, which will prevent the shavings from becoming damp and sticking to the plumage, which might discolor it. The cage is left near enough to the fire to keep the bird warm while drying. If the temperature is too low, the bird will shiver, when he should be moved nearer. If too high, the bird will pant, when he should be moved further away. This should proceed at a regular and fairly rapid rate. If left too near the fire, the drying may be too rapid and this may cause the feathers to curl. Many conditioners help the drying of the tail and other portions of the plumage which tend to dry slowly by fanning them. This also helps in the fluffing out of the feathers and gives a fine finish to the birds. An electric fan is very useful for this purpose. While fanning, the feathers should be lifted and spread, to hasten drying.

When the bird is dried, he should gradually be accustomed to a lower temperature, so that he will not feel the change so greatly when shipped and will not be so likely to catch cold. After washing, the birds should be handled as little as possible, and the litter used in the coop or pen kept clean by changing as often as necessary, so as not to soil the plumage.

Bluing white birds.—It used to be common to blue white birds after washing. This was accomplished by adding bluing to the final rinsing water in about the amount usually used for bluing clothes. Many birds are spoiled, however, by the use of too much bluing and also by getting a streaky effect, so that the custom is not nearly as prevalent as formerly.

Bleaching and cleaning.—It is very common practice to bleach and cleanse the plumage of white birds after washing by the use of peroxide. This should never be attempted except in the case of birds that are entirely white. Technically, this may be held to be faking, but actually it is so commonly employed that it can hardly be so considered, and every conditioner of white birds should know how it is done. This process is supposed to bleach out creaminess to some extent and is also valuable in removing stains and cleansing the plumage. It will not take out brassiness.

The time to use the peroxide is when the feathers on top of the head begin to fluff out. Warm peroxide is more effective than cold, and it can readily be warmed by removing the stopper of the bottle and standing the bottle in a pail of warm water. Pour the peroxide into a large, shallow earthenware dish, such as the old-fashioned washbowl. Lay the bird on his back in the peroxide and press the wings and tail down in the fluid. Turn the bird on each side likewise, and be sure that the peroxide comes in contact with all parts of the plumage, but particularly that of the hackle, back, wings, shoulders, saddle and tail. A sponge is also helpful in applying the peroxide, as the liquid can be taken up in the sponge and squeezed out at the point desired so as to run over the plumage. Place the bird back in the cage and allow to dry.

Other care of plumage.—In plumage where a high gloss or sheen is desired, rubbing with a silk cloth will enhance this. It is said that the appearance of the plumage of a black bird can be improved by rubbing with silk dipped in vinegar. Rubbing the plumage of black or red birds with a flannel cloth dampened with cocoanut oil will impart to it a very high gloss. Do not, however, use so much oil as to make the plumage oily. The plumage should be carefully gone over for the purpose of plucking any off-colored body feathers, such as black feathers in Barred Rocks. In addition to plucking the off-colored body feathers, it is common practice to pluck a few of the feathers in any body section which shows badly defective markings, and which in consequence mar the excellence of the section. For example, a few feathers showing badly broken barring, plucked from the hackle of a Barred Plymouth Rock, or a few feathers showing black lacing, plucked from the hackle of a Rhode Island Red male, may greatly improve the general excellence of that section. Care must be taken not to pluck too many feathers, as this practice carried to excess destroys the smoothness of feathering and greatly injures the appearance of a fowl. The main wing or tail feathers are seldom plucked, as their absence is rather heavily penalized or may even disqualify a bird. It is therefore essential to select birds for exhibition which are pretty sound in wing and tail.

Not infrequently the tail feathers may grow too long to make a nicely balanced tail. In other cases, the tail feathers may be broken, or otherwise injured. They may be pulled and new feathers allowed to grow in. About six weeks are usually allowed for the growth of such new feathers.

Shipping birds to the show.—It is desirable to have the birds arrive as short a time before the show opens as possible. Otherwise the birds become tired out and do not show to as good advantage. Therefore ship as late as is safe. Shipment is by express, of course, and should be fully prepaid. During shipment the birds must be comfortable. This means that the coops must be large enough to allow the birds to stand erect and to stand in a natural position. If the coop is too small, a bird may greatly injure its tail by constantly hitting or rubbing it against the sides of the coop. A coop 12 inches wide, 22 inches long and 25 inches high is of suitable size. For females a coop may be used, if desired, of the same dimensions, except that it is only 18 inches high.

Coops are usually made of light wood, but are sometimes used which are made of pasteboard, or which have cloth sides. The sides are tight, to protect the fowls from drafts.

In mild weather the top of the coop may be merely slatted, but in winter it should have a wooden or muslin top. Where wooden tops are used, ventilation holes in the sides of the coop near the top are necessary to prevent the birds from suffocating. Where muslin is used, a space of about an inch should be left on each side between the muslin and the side of the coop for ventilation. Only a single bird should be shipped in a coop or compartment. The shipping coop should be constructed so that it can be securely fastened shut, but at the same time is easy to open. This is especially important when one is unable to attend the show to uncrate the birds, for coops not so constructed are often so badly damaged in getting them open that they are not suitable or safe to use in returning the birds from the show.

The shipping coop must be clean. If it is not, the plumage will be soiled. Wipe out the inside of the coop carefully to make sure that no dust has accumulated. Clean dry shavings should be placed in the bottom of the coop to absorb the moisture from the droppings, and thus help to keep the fowl clean. If the shipment is short, and especially if the owner is to attend the show and see to the uncooping and feeding of his birds, it is best to omit any feed. Otherwise a small amount can be put in the coop, never a large supply. No water should be provided unless the fowl is to be two days or more on the road. Water in a coop will surely be slopped, and this will make the shavings damp and may soil the bird's plumage. A mangel, or something similar, placed in the coop, will supply the bird with necessary moisture and is preferable to water. Every detail should be carefully attended to which will help to put the bird in the show room in the best possible condition. The good results of a lot of time and effort spent in conditioning birds may be wholly or partially nullified by a little carelessness or oversight.

At best, and particularly in winter, there is danger of the birds catching cold. This is enhanced if the coops are placed next the steam pipes in the express cars, where the birds become overheated. To avoid this, if possible, tack a notice on the coop asking the express company not to put the coops in such a place. The coops should, of course, be plainly and fully addressed, and the owner's name and address should likewise be stenciled on them.

It is by all means desirable to travel on the same train with the birds, if possible, particularly if any transfers are necessary. In this way a person may often prevent delays and also see that the coops are properly placed in the cars.

Care of fowls in the show.—It is by all means desirable to go to the show to do one's own uncrating, and cooping, and to care for the fowls. The show management does the best that it can, but has so many birds to see to that it is impossible for it to give the care and attention that the owner can if he is on the ground. After locating the shipping coops and the show coops which are assigned to the birds, the first thing to do is to effect the transfer of the birds from the former to the latter. Wipe out the coops to make sure they are clean before putting the birds in. It will often save sickness to clean and disinfect the drinking and feed cups before letting the fowls use them.

Often birds which have seemed to be tame will become confused and frightened at the noise and movement in the show room, and may jump against the top of the coop in an effort to get out. This is likely to injure the comb. A newspaper or other covering, placed on top of the coop, will stop this, as the birds see no chance of escape in that direction.

It is desirable for the exhibitor to lock the coops containing his birds. This will prevent individuals from taking the birds out of the coops and handling them, a practice which may lead to injury. It may also prevent the loss of birds, as thefts occasionally occur.

Locking the coops will necessitate tending to the feeding of the birds, but most exhibitors prefer to do this anyway.

Do not overfeed. Two light feeds of grain a day are sufficient. There is more danger of overfeeding than of under-
feeding. A little green feed may also be given. A small
piece of apple, turnip, or other vegetable once in two days
is sufficient. A bit of raw meat, such as hamburg steak, may
be fed as well, if desired. Water should be given three
times a day and left so that they can drink for about half
an hour. Then it is best to take it away, as it is likely to be
spilled and will make the coop damp and dirty. A fresh
supply of shavings should be put in the coop whenever they
become scanty or dirty.

As nearly immediately preceding the judging as possible,
the head, comb, face and wattles should be wiped off with
a mixture composed of one-half alcohol and one-half sweet
oil. This will help to brighten up these parts. If they are
quite pale, massaging with the fingers will also be found
beneficial. At the same time the shanks and toes should be
wiped with the same mixture to brighten them.

Covering the top of the coop at night will darken it and
enable the fowls to sleep better. It may also protect the
fowls from marked changes in temperature, if the heat is
allowed to go down in the hall.

Treatment of birds after the show.—After the show
closes, it is best to attend to the cooping of your own birds
and to see that they reach the express company representa-
tives' hands. This will insure prompt shipment. Be sure
that your coops are properly labeled for their return jour-
ney. Return shipment should, of course, be made by the
same express company in order to take advantage of any
lower rate.

After the birds arrive home, it is well to keep them
separate from the rest of the fowls for a few days, to make
sure that no disease develops before they are put in the
breeding pens.

Making entries.—Make entries as early as possible, so
as to avoid the danger of their not arriving before the entries

close or of their being refused on account of lack of room. It is very desirable to make more than a single entry in a class. Often the owner's judgment may be faulty and the bird he selects as the best might not win, while if he entered two or three, one of the others might do so. Also, judges have individuality in their work, and when competition is close a bird might be placed under one judge, while another would be placed instead under another judge. Therefore, the making of more than one entry where the birds are nearly equal in quality betters the exhibitor's chance to win.

In making several entries, it is well to make them at two or more separate times. The entry numbers, and consequently the coops, are assigned in the order the entries are received, so that by making the entries at different times position will be secured in different locations in the class. This may prove to be of considerable importance, as the light may be better in one part of the class than in another. In shows where the coops are double tiered, if the entries for a class are made at different times, at least two should be made together each time, to insure one of the birds being placed in the upper tier of coops, where the light is better.

INDEX